Powder Metallurgy Diamond Tools

Powder Metallurgy Diamond Tools

Janusz Konstanty
Faculty of Metallurgy and Materials Science,
AGH University of Science and Technology,
Mickiewicz Avenue 30,
30-059 Krakow,
Poland

2005

ELSEVIER

Amsterdam – Boston – Heidelberg – London – New York – Oxford
Paris – San Diego – San Francisco – Singapore – Sydney – Tokyo

ELSEVIER B.V.
Radarweg 29
P.O. Box 211, 1000 AE Amsterdam
The Netherlands

ELSEVIER Inc.
525 B Street, Suite 1900
San Diego, CA 92101-4495
USA

ELSEVIER Ltd
The Boulevard, Langford Lane
Kidlington, Oxford OX5 1GB
UK

ELSEVIER Ltd
84 Theobalds Road
London WC1X 8RR
UK

First edition 2005

ISBN: 1-856-17440-9

∞ The paper used in this publication meets the requirements of ANSI/NISO Z39.48-1992 (Permanence of Paper).
Printed in Great Britain.

I would like to dedicate this book to my parents

Contents

Acknowledgements

I am deeply grateful for invaluable assistance that I have received at various stages of this project from numerous companies and institutions, including:

Asahi Diamond Industrial Australia Pty Ltd. (Mona Vale, Australia)
BASF Polska Sp. z o.o. (Katowice, Poland)
Bodycote Hot Isostatic Pressing n.v. (Sint-Niklaas, Belgium)
Breton S.p.A. (Treviso, Italy)
Diatech S.C. (Warsaw, Poland)
ECKA Granulate GmbH (Fürth, Germany)
EHWA Diamond Ind. Co. Ltd. (Osan, Korea)
Element Six (Shannon, Ireland)
EPSI N.V. (Temse, Belgium)
Eurotungstene Metal Powders (Grenoble, France)
Foxmet S.A. (Niederdonven, Luxemburg)
GE Superabrasives Europe (Dintikon, Germany)
H.C. Starck GmbH (Goslar, Germany)
Hilti Corporation (Schaan, Liechtenstein)
Höganäs AB (Höganäs, Sweden)
Kennametal Inc. (Latrobe, U.S.A.)
Marian Kaźmierczyk Narzędzia Diamentowe (Krakow, Poland)
MC DIAM Sp. z o.o. (Warsaw, Poland)
OMG (Kokkola, Finland)
Sanxin Wire Die, Inc. (Charlottesville, U.S.A.)
Syndimant GmbH (Hannau, Germany)
The Cobalt Development Institute (Guildford, England)
Umicore (Olen, Belgium)

as well as from countless individuals worldwide.

Additionally, I wish to gratefully acknowledge Mrs. Joanna Borowiecka-Jamrozek of Kielce University of Technology for her able assistance with the SEM.

Special thanks are due to Professor Hanna Frydrych of University of Science & Technology, Krakow, for encouragement and financial support.

CHAPTER 1

Introduction

Diamond, an allotropic form of carbon, is the hardest mineral known to man. The flawless, or virtually flawless, pieces of diamond, when cut and polished, have always been valued as the most precious gems. Except rarity of gemstones and their decorative merits, there are also many other properties which make diamond a unique material [1]. It has the highest thermal conductivity at room temperature, the highest bulk modulus and the highest critical tensile stress for cleavage, an extremely high thermal conductivity, very low coefficients of friction and thermal expansion, and it is relatively inert to chemical attack by common acids and bases.

Until now, only few of these properties have fully been exploited in advanced industrial applications due to limited size and high price of diamond. The latest progress in diamond synthesis, however, is starting to remove the existing barriers, opening new opportunities for diamond as an advanced engineering material.

1.1 Historical Development of Diamond Tools

The modern application of diamond tools is roughly a century old although the early use of diamond as an engraving tool goes back to 350 BC [2]. In Christ's time, splinters from broken diamonds were first applied as a set in iron handles, i.e. metal-bonded diamond tools [2,3].

The next milestone in the history of diamond tools was to take place in 1819, when the first patent for a diamond wire-drawing die was granted to Brockendon in England. At that time, however, it proved impossible to implement this invention into practical use and it took around 40 years until the first diamond wire-drawing die was successfully made and utilised by Milan and Balloffet in France [2].

Meanwhile, in 1824, Pritchard started to use shaped diamond wheels to grind and polish microscope lenses [2–4]. These wheels were made by hammering diamond grits, of adequate fineness, into the surface of a cast iron body [3].

In 1854 a French engineer Hermann applied for a patent for a single-crystal diamond tool for cutting, turning and shaping hard stones which, upon improvements made a year later, was converted into a tool with multiple diamonds [2]. A few years later, in 1862, Leschot of Geneva was granted a patent covering a complete drilling rig which was to find practical application on a broader scale more than a century after the first description of a primitive diamond rock drill had appeared in Diderot's Encyclopaedia [2,4].

The first diamond circular sawblades for cutting stone were developed by Fromholt in France in 1885. Thirteen years later, a large diameter blade was first used in practice in the Euville stone quarries. The early blades used Brazilian carbonado diamonds set around their periphery. Carbonado was a valued material at that time [4] because, being a cryptocrystalline mass of small crystals locked in random directions, it was strong and resistant to cleavage. Such carbonado blades were utilised to cut limestone and marble during the construction of large buildings in Paris in the 1900s [2].

Further progress in the tool production routes took place in the period between 1927 and 1931, when the earliest patents describing the manufacture of metal matrix abrasive tools, by making them from powders, were issued in the USA and Great Britain. According to Gauthier (1927) [5], the powder mix was to be consolidated by cold pressing only, whereas Neven (1931) was probably the first to suggest hot pressing [6]. The first metal powder used was electrolytic iron.

The idea of bonding diamond by means of metal powders dates back to 1883, when Gay described the manufacture of abrasive materials by incorporating traditional abrasives, such as quartz or emery, in a metal matrix [6]. He mentioned the use of brass, iron or steel powders and proposed to make a good use of powder metallurgy techniques, such as hot pressing or infiltration, to form the matrix [6]. It had taken, however, a few decades until the inventions of the 1920s and 1930s made refinements to the Gay's ideas and apparently speeded up the development of diamond grit impregnated tools which were finally introduced to industrial application around 1940 [2].

Bonds other than metal were also being developed during this period. In 1925 the Bakelite Corporation took out a patent on the first phenolic resin bond [7]. In the early 1930s resin-bonded wheels, containing 'fragmented' natural diamond grit, were patented by Wickman Ltd. in England (1933) [2], Voegeli & Wirz in Switzerland (1934) [2], and Norton Co. in the USA (1934) [8].

Until the early 1950s the developments in diamond tools were relatively slow. In that period only mined diamond crystals were available. These were formed millions of years ago, under conditions of intense heat and pressure acting on the carbon, and later ejected to the surface by volcanic eruptions.

Much faster developments in the tool manufacturing technology, which have been seen over the last 50 years or so, may chiefly be attributed to the 'invention' of synthetic diamond. Efforts to manufacture synthetic diamond crystals date back at least several hundred years. They had remained fruitless until 1953, when positive and fully reproducible results were obtained by a team of researchers at ASEA [9]. Quite independently, and entirely without knowledge of what ASEA had been doing, General Electric announced its capability to manufacture synthetic diamonds on an industrial scale in 1955 [10]. While ASEA kept the diamond experiments secretive, GE was first to describe the process in the scientific literature [11] and patented it [12].

Permanent progress in the manufacturing technologies fostered the commercial importance of synthetics, which is now accounting for over 95% of all industrial diamonds consumed [13]. It is worthwhile to mention that the last five decades witnessed a spectacular, 50-fold or so, increase in the total consumption of industrial diamond. Over this time modern production techniques based on diamond tooling have been implemented into evolving areas of industrial activity enabling to do the job faster, more accurately and at less cost. They revolutionised machinery and processing techniques in the stone and construction industries, road repair, petroleum exploration, woodworking, cutting frozen foods, production of various parts and components made of glass, ceramics, metals, plastic and rubber, etc.

The milestone progressions and developments, which followed commercialisation of synthetic diamond on a broader scale, can be listed chronologically as follows:

1960s: Metal-clad diamond was developed for application in resin bonds, which coincided with the introduction of polyimides by Du Point [14]. Wire saws for sawing stone were produced in Italy in 1969 [15]. They contained diamond grit embedded in an electrodeposited metallic matrix. Cubic boron nitride (CBN) was introduced to the industry in 1969 to complement diamond in machining ferrous alloys [16].

1970s: Synthesis of high-quality 'saw' diamond was developed for demanding stone working applications such as sawing granite [17]. Polycrystalline diamond (PCD) became available on a broader scale and made extensive inroads into applications which had been the domain of cemented carbides [17,18].

1980s: Coated 'saw' grits were introduced into broader application [19]. A new class of phenol-aralkyl thermosetting resins, offering improved tool performance, was developed for application in resin-bonded diamond and CBN grinding wheels [14].

1990s: Major breakthroughs in low-pressure synthesis of polycrystalline diamond by chemical vapour deposition (CVD) were achieved. This resulted in commercialisation of CVD diamond coated cutting tool inserts [20], twist drills, and 'free-standing' thick CVD diamond films [18] brazeable to the tool support.

In the new millennium the market for diamond tools continues to grow rapidly. The most recent figures [19] indicate that the demand for diamond abrasives reached an impressive volume of 1 billion carats in 2000, as compared with approximately 380 million carats in 1990 and 100 million carats in 1980. The current trend is to diversify into applications still dominated by traditional abrasives with particular interest in developing linear blades for sawing granite as well as in applying diamond grits on a broader scale in the surface finishing operations [21]. Nowadays the rapid diamond price decline [22] makes industrial diamond a commoditised product capable of competing, in terms of its price/performance ratio, with conventional abrasives such as silicon carbide and aluminium oxide.

1.2 Classification of Diamond Tools

The term 'diamond tools' has a very broad meaning. The existing classifications of diamond tools are based on various criteria, such as the quantity of diamond involved and its origin, outward appearance and internal structure of the tool, its application, etc.

For the purpose of this book it is convenient to arrange the tool types into categories which would be distinctive with respect to the various manufacturing methods involved. Such a classification is shown in Fig. 1.1.

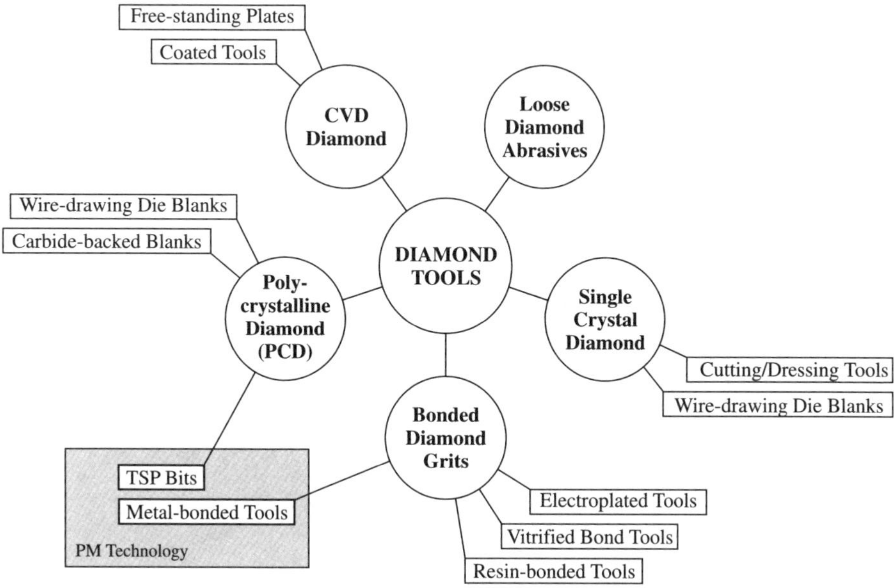

Figure 1.1 Classification of diamond tools.

1.2.1 Loose Diamond Abrasives

A loose diamond micron powder constitutes the simplest diamond tool (Fig. 1.2). The existing diamond sizing standards define micron powders as nominally finer than around 84 μm although various sizing criteria and techniques are utilised [23]. Both natural and synthetic micron abrasives, mostly in a paste or liquid suspension form, are used for a wide variety of fine grinding and polishing operations. The most typical applications are preparation of metallographic, ceramic and mineralogical specimens [24], finishing of diamond cutting tools [25], profiling and calibration of diamond wire-drawing dies [26], polishing gemstones, sizing and finishing hardened steel and tungsten carbide tool components [27], etc.

1.2.2 Single-Crystal Diamond

Single-crystal diamond tools, both natural and synthetic, are used as cutting tools, dressers and wire-drawing dies. Poor availability of large natural stones, variations in their quality, and time-consuming diamond selection and tool tip preparation procedures have, however, markedly restricted the consumption of natural stones over the last decade or so. The recent developments in high pressure–high temperature diamond technology make it possible to produce synthetic single-crystal diamonds having sufficient size and consistency of shape, crystallographic orientation and performance not achievable with natural products [18,28,29].

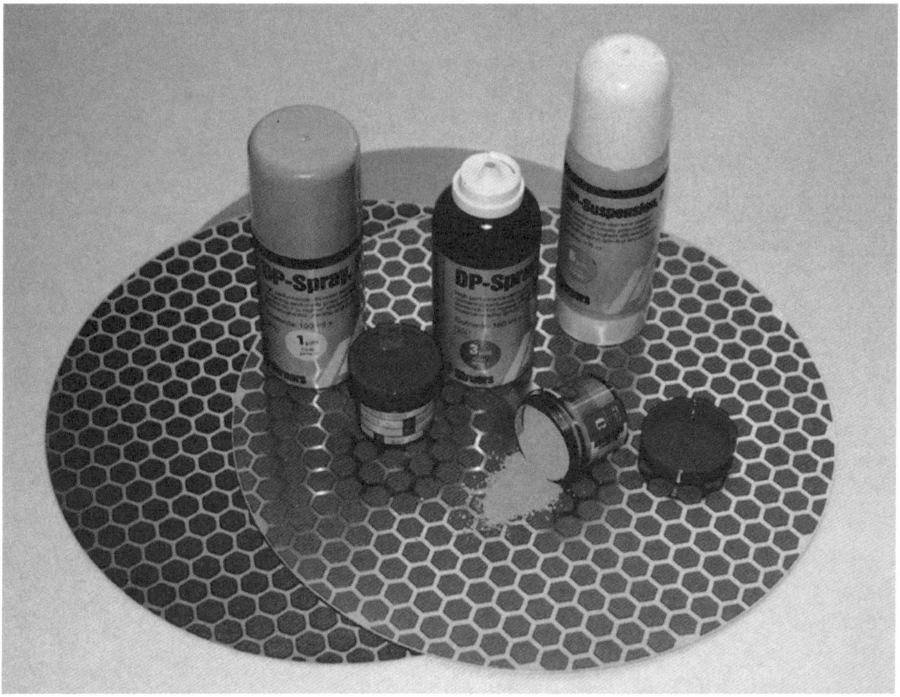

Figure 1.2 Loose diamond micron powder products.

1.2.2.1 Cutting/dressing tools

Diamond cutting blanks and dresser logs are produced by laser cutting of raw monocrystals [30]. Synthetic diamonds are more regular in shape than natural stones and the crystallographic orientations of each crystal are easy to define. Therefore large synthetic crystals can readily be cleaved into plates and laser cut [28] to produce suitable shapes in a broad range of sizes, up to around 8 mm as standard [18,30], with crystallographically orientated sides and cutting edges as shown schematically in Fig. 1.3.

Because diamond is an anisotrophic material, its rate of wear varies with the crystallographic orientation. For instance, for the planes of the form {100} the directions exhibiting the highest and the lowest resistance to wear are ⟨110⟩ and ⟨100⟩, respectively [31]. This becomes a matter of practical concern in the case of synthetic diamond dresser logs where it is easy to affect the wear resistance of the tool by rotating the blank about its longitudinal axis to use the most wear-resistant crystallographic directions as the working directions.

The single-crystal diamond tool blanks have long been applied for high-precision machining of non-ferrous metals [28,30,32], unsintered carbide blanks [32], plastics and glass [28,30], wood and laminates [25], dressing of profile grinding wheels

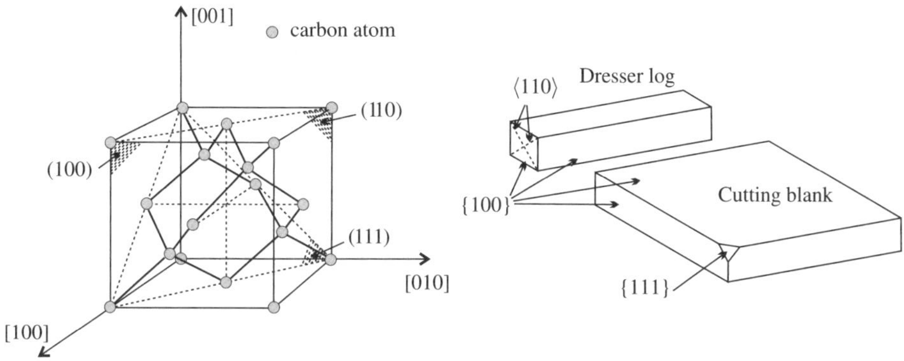

Figure 1.3 Crystal structure of diamond (left) and idealised shapes of single-crystal diamond tool blanks (right).

[28–30], etc. Nowadays the synthetic diamond tools are used on much broader scale due to the ease of fabrication, improved brazeability onto the tool holder, extended tool life, and reduced machine downtime for the repolishing/resharpening operations [30]. They are also gaining ground in applications where natural diamonds were traditionally used, such as in hardness testing indentors, scratching and scoring styli, small engraving tools, as well as in measuring instruments [30].

1.2.2.2 Wire-drawing dies

In performance terms natural diamond has long been valued as an ideal die material. The main drawback of natural stones is their high cost since only high quality crystals are entirely suitable for the die fabrication process which includes laborious inspection and flatting operations [33]. The advantage of synthetic monocrystalline diamond blanks is that the opposing faces of each blank are easily produced by cleaving a suitably grown crystal on the octahedral {111} plane [28]. From that point on, the die fabrication process is virtually identical to that of natural diamond dies. Initially, the die blank is mounted in a suitably shaped casing with carefully selected powdered filler metals [28,34] by means of high-frequency induction heating and under moderate pressure. After laser piercing, the die opening is profiled and polished ultrasonically with a shaped steel needles and abrasive diamond slurry. The final operation consists in calibrating the bearing zone of the die with oscillating wire which pulls diamond slurry through the bearing area.

These days high-grade synthetic single-crystal die blanks are being used in increasing quantities displacing natural diamond dies. Synthetic stones have low and controlled level of metallic inclusions, which allows superior fracture resistance [28], and are capable of withstanding temperatures exceeding 1000°C in a non-oxidising environment. By synthesising stones suitable for fabrication of die blanks with a predefined crystallographic orientation it is possible to eliminate the problem of uneven die wear which frequently occurs with natural diamond dies [33].

Single-crystal diamond dies find increasing application in drawing ferrous alloys, non-ferrous metals and alloys, precious metals, and high temperature drawing of refractory metals, offering unrivalled productivity and quality of the finest sizes of wire [33].

1.2.3 Bonded Diamond Grits and Powders

This is by far the largest group of diamond tools used in a variety of workpiece machining applications. They consist of diamond particles, either mono- or polycrystalline, which are embedded in diverse metallic or non-metallic matrices by various fabrication routes.

1.2.3.1 Electroplated tools

One of the earliest applications of metal-bonded diamond tools manufactured by electro-deposition of the working layer onto a metallic tool body was in dentistry [35,36]. Nowadays diamond electroplated tools are used on a broader scale for sawing, grinding, boring, profiling, brushing, and surface-finishing of glass, porcelain, cemented carbides, refractories, plastics, and other materials.

The formation of the diamond-impregnated layer starts with bringing diamond grit, or powder, into contact with the working face of a rigid, or flexible, metal body whereas the rest of the tool is masked off with lacquer [37] or wax [38]. The whole assembly is then submerged in an electrolyte where a coating of a suitable metal matrix is electro-deposited until its thickness is sufficient to hold one layer securely in place, or occasionally more than one layer, of diamond particles. The metal most frequently applied is nickel [37] but nickel alloys, such as Ni–Co [39,40], Ni–Mn or Ni–Co–Mn, may also be used [39] by simultaneous deposition of the required metals from mostly their sulphate solutions, with various additives, under strictly controlled conditions. For better adhesion with the substrate, the tool body is often pre-coated in the absence of diamond [40] (Fig. 1.4).

Both natural and synthetic diamond grits, and powders, are used for electroplating. Synthetic crystals often require purification prior to electro-deposition since metallic inclusions, originating from the metal solvents used during diamond synthesis, may promote harmful overplating of the abrasive particles [41]. Natural diamonds, being inherently free from metallic inclusions, do not require any chemical or thermo-chemical pre-treatments and are perfectly suited for the tool fabrication by electroplating.

1.2.3.2 Vitrified bond tools

A characteristic feature of vitrified bond diamond tools is high porosity which amounts up to 55 vol% of the diamond-impregnated layer [37,42]. The pores play an important role since they provide clearance chambers for storing and removing the detritus generated during grinding. Another advantage of high porosity is good distribution of the coolant throughout the tool–workpiece contact area.

Compared to the other types of diamond-impregnated tools, vitrified bond diamond tools have found only limited application [43]. The unique properties of

Figure 1.4 Electroplated tools. (Courtesy of EHWA Diamond Ind. Co., Ltd.)

vitrified matrices, however, such as a relatively long tool life and dimensional accuracy, ease of dressing, and ability to work at elevated temperatures when combined with a friable diamond grit ensures that the tools remain sharp during use with the minimum of re-dressing [44]. This makes them useful in grinding single-crystal and PCD diamond tools [42], cemented carbides and ceramics [45] as well as certain composite materials [37].

The main compounds used to produce vitrified bonds are SiO_2, Na_2O and CaO that comprise the basic glass [42]. Some other constituents, such as SiC, Al_2O_3, B_2O_3, ZnO, K_2O, as well as feldspar, clays and other natural minerals, are also added in varying proportion to modify the characteristics of the material [37,42]. The powdered bond-diamond abrasive mix, with diamond concentration typically between 75 and 150 (18.75–37.5 vol%) [46], is either cold pressed and fired in a furnace at 900–950°C [47] in an inert atmosphere or, preferably, hot pressed at around 730°C [42]. The advantage of the latter route is markedly shorter processing time at lower temperature. Such a procedure yields tools with lower porosity and is less detrimental to the diamond toughness if a synthetic abrasive has been chosen (Fig. 1.5).

1.2.3.3 Resin-bonded tools

In the vast majority of cases resin bonds are based on either phenolic resins or, more thermally stable, phenol-aralkyl resins [14]. They are used in the form of a very fine powder which accounts from around 30 to 40 wt.% of the bond in dry and wet

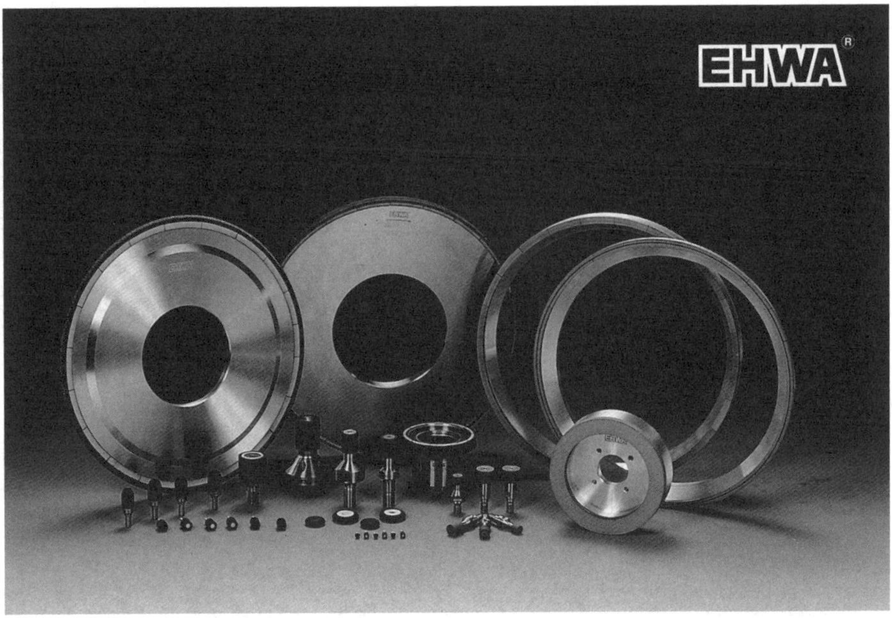

Figure 1.5 Selection of vitrified bond tools. (Courtesy of EHWA Diamond Ind. Co., Ltd.)

grinding wheels, respectively. To impart strength to the bond, the resins are mixed with SiC filler [37]. Various oxide-base drying agents, graphite or polytetrafluoroethylene lubricants, as well as aluminium, copper or silver powders are also added in small amounts in order to reduce and dissipate the frictional heat generated in grinding [14,37]. The bond components are carefully mixed with a suitable amount of natural or friable synthetic diamond grit. The mix is poured into a mould and hot pressed to shape the abrasive rim which is subsequently attached to the tool body by using a suitable adhesive [14]. Optionally, the mix may be hot pressed directly onto the tool hub which has undercuts on its circumference in order to aid mechanical keying. For perfect bonding a thin film of heat-resistant adhesive is usually applied [14].

Depending on the wheel application the diamond concentration may range from around 50 to 125 (12.5–31.25 vol%) [48]. Most resin-bonded diamond tools (Fig. 1.6) take advantage of metal-clad abrasives which possess improved retention and heat dissipation characteristics as compared with unclad diamond [45]. The metal cladding options are nickel-base alloys, copper or silver. Nickel-clad diamonds, containing 30–60 wt.% nickel alloy, may have a rough, 'spiky' surface which aid diamond retention in the bond under severe grinding conditions [48,49]. Copper-clad diamond, containing 50 wt.% copper is mainly dedicated to dry grinding applications. Silver-clad diamond, containing 50 wt.% silver, has recently been developed for cemented carbide grinding with straight oil coolant [50,51]. As in certain nickel-clad grades, the spiky silver coating formation enhances bond retention. It has also

Figure 1.6 Resin-bonded tools. (Courtesy of EHWA Diamond Ind. Co., Ltd.)

been found to reduce friction between the tool and the workpiece [51]. By increasing thermal conductivity of the working rim of the grinding wheel both silver and copper perfectly improve heat dissipation, thus reducing the severity of the heat pulse to the generally temperature-sensitive resin bonds [45,51].

Resin-bonded diamond tools offer cost-effective grinding where close dimensional tolerances and good surface finish are critical. For that reason they have found application in grinding of cemented carbides [37,52], oxide and non-oxide ceramics [53,54], glass [54], as well as in fine grinding and polishing of natural stones [55].

1.2.3.4 Metal-bonded tools

In the production volume terms, metal-bonded tools account for around two-thirds of the whole bonded-grit diamond tool market [43]. In the metallurgical sense, the term 'metal bond' should be applied to electroplated products rather than to metallic-matrix diamond composites which, after all, most frequently occur in the colloquial technical language as 'metal-bonded' or 'sintered' tools. As characterised by the latter synonym, the distinctive feature of what, for the purpose of this book, has been classified as 'metal-bonded diamond tools' (Fig. 1.7) is an application of different powder metallurgy (PM) techniques to the manufacture of the diamond-impregnated working layer.

The various PM fabrication routes as well as the most important application issues for metal-bonded diamond tools are dealt with in detail in the next chapters of this book.

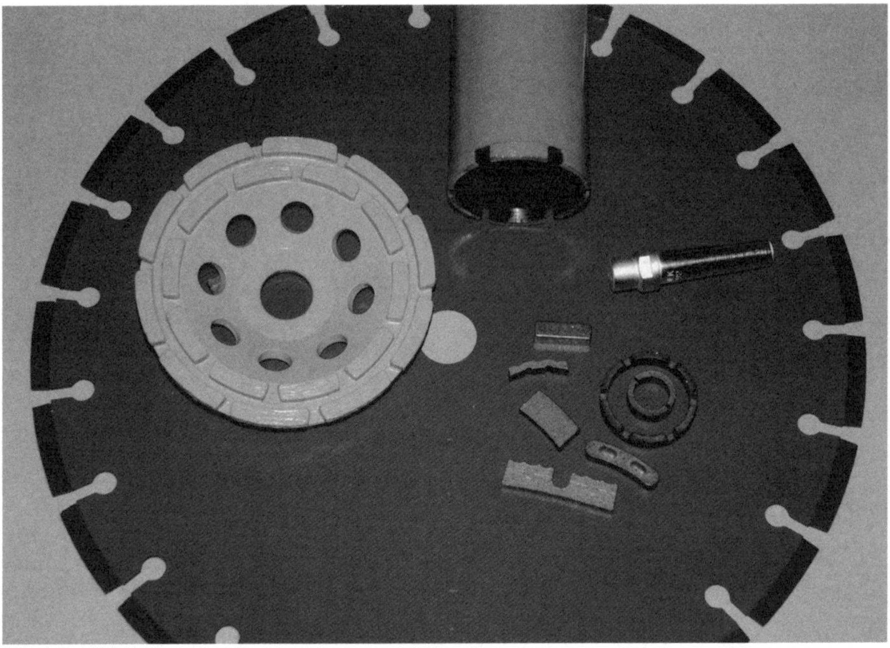

Figure 1.7 Metal-bonded tools and diamond-impregnated tool components.

1.2.4 Polycrystalline Diamond (PCD)

Since the early 1970s, PCD has become established as a high-performance and cost-effective alternative to conventional tools in the most demanding industrial applications, such as high-temperature drawing of wire [34], machining of metal matrix composites [56], fibre-reinforced plastics, high SiAl alloys [18], and abrasive wood composites [57], drilling in various rock formations [58], etc.

PCD is produced by consolidating together accurately sized, high-quality diamond powder in the presence of either metallic or non-metallic binding phase at temperatures between 1200 and 1600°C and at high pressures of the order of 6 GPa [59]. When cobalt is used as the solvent/catalyst metal, it promotes the conversion of graphite, which has formed at exposed grain surfaces, into diamond through a dissolution–reprecipitation process under conditions of diamond thermodynamic stability. This leads to a marked amount of intergrowth between the randomly orientated diamond grains [59]. The final grain size depends on the average grain size of the diamond starting powder which, in commercially available grades, varies from 2 to 50 μm [59,60]. As would be expected, fine-grained PCD grades are tougher and enable better surface finish to be achieved, whereas the coarse-grained grades are more resistant to wear.

The presence of around 8 vol% cobalt in the commercial grades of PCD impairs the mechanical properties of the material after its prolonged exposure to temperature in excess of 700°C. There are two reasons for this. First, the presence of cobalt promotes diamond graphitisation and, second, the large difference between the thermal expansion coefficients of diamond and cobalt leads to high internal stresses at elevated temperature [34,61].

In order to impart better thermal stability to the material it is necessary to either leach out the residual cobalt from the PCD composite or to replace the metallic phase with a suitable non-metallic binder [62,63]. When silicon is used, instead of cobalt, a catalytically inactive SiC binder is created by the reaction between silicon and diamond. Since the thermal expansion coefficient of SiC is very close to that of diamond, the PCD grades which contain SiC, typically around 19 vol% [59], can withstand processing at temperatures up to 1200°C in an inert or reducing atmosphere [59,62].

1.2.4.1 Wire-drawing dies

The smaller PCD wire-drawing die blanks are produced as free-standing cylinders or hexagons, whereas the larger cylindrical blanks are bonded to a supporting tungsten carbide ring [34,60]. The die manufacture is essentially the same as in the case of single-crystal dies with an additional possibility of piercing the die blank by electric discharge machining (EDM) if there is sufficient amount of cobalt in the PCD mass to render it electrically conductive [26,64].

The PCD dies are currently manufactured as either containing cobalt or thermally stable grades [60,64]. Both products are ideally suited to economic production of wire on multi-line machines, where fast drawing speeds, die life and product quality are crucial. Unlike single-crystals, PCD dies are available in the full range of sizes and therefore they can be used to draw wire from 20 mm diameter down to micron sizes [64] (Fig. 1.8).

1.2.4.2 Carbide-backed blanks

PCD turning, reaming, boring, sawing and milling tools are produced exclusively as cemented carbide-backed blanks [65,66]. For rock drilling applications the carbide-backed PCD blanks are used as a complement to the thermally stable unbacked material [67,68]. The PCD layer is integrally bonded to the cemented tungsten carbide substrate at ultrahigh pressure and high temperature in the presence of cobalt.

The commercial products are available as discs, of diameter up to 74 mm, or in a variety of shapes and sizes suitable for direct tipping cutting tools. The individual tool inserts are made from PCD discs by cutting with wire EDM [69] or laser [70] and edge grinding with either mechanical means or spark erosion (rotary EDM) [69]. These inserts are subsequently induction brazed or mechanically clamped to the desired tool body (Fig. 1.9).

1.2.4.3 Thermally stable PCD (TSP) bits

With the advent of thermally stable PCD, in the early 1980s [62], it has become possible to produce TSP bits and TSP hybrid bits (Fig. 1.10) by means of PM technology. These products are widely used in mining and geological exploration drilling to

Figure 1.8 PCD wire-drawing dies. (Courtesy of Sanxin Wire Die, Inc.)

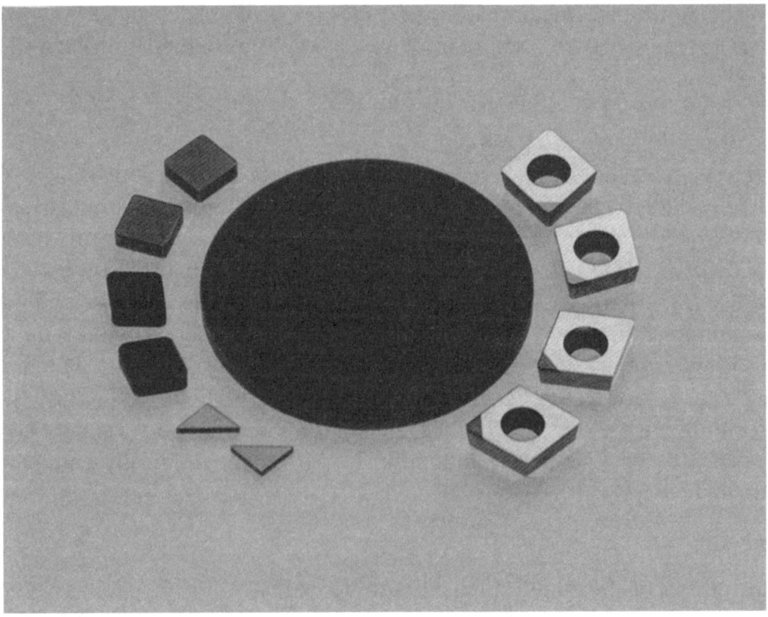

Figure 1.9 Selection of PCD cutting tools and semi-finished products.

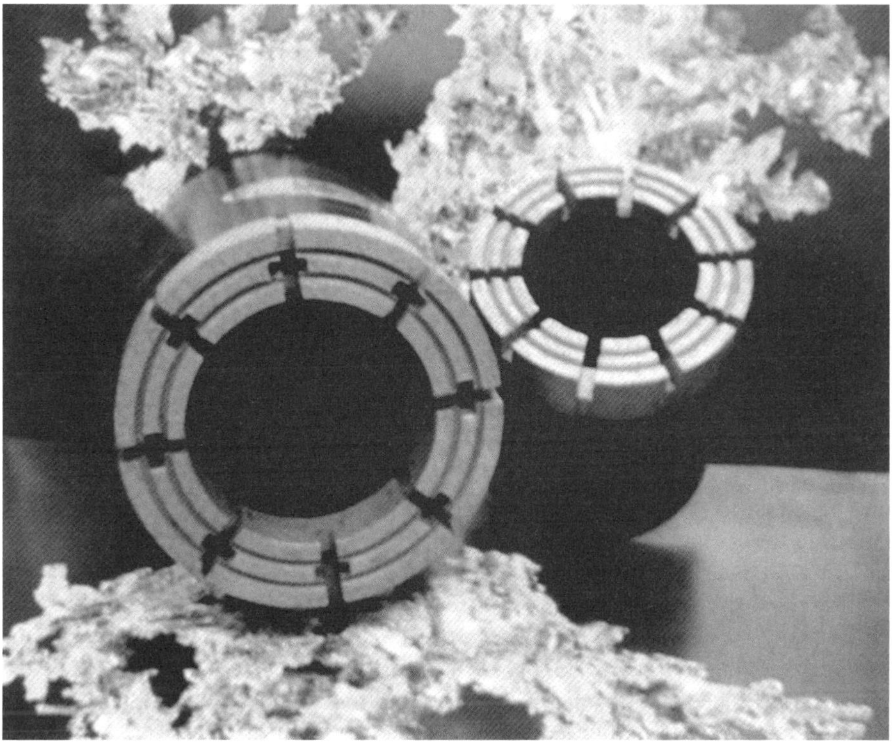

Figure 1.10 Hybrid core bits. (Courtesy of Asahi Diamond.)

complement the other drill types, such as impregnated diamond bits, surface-set diamond bits, carbide-backed PCD bits as well as conventional cemented carbide insert bits [71].

A detailed account of TSP bits PM fabrication by means of the infiltration technique is given in Chapter 4.

1.2.5 CVD Diamond

An alternative route to synthesise diamond is by various chemical vapour deposition (CVD) techniques, which consist in deposition of metastable polycrystalline diamond film from a gaseous phase onto a suitable substrate maintained at between 700 and 1000°C [17,18,72]. In general, a mixture of carbon carrier gas, typically 1% methane, and hydrogen is introduced into a CVD reactor where the gases are decomposed, below atmospheric pressure, by utilising various energy sources, such as the electric arc, hot filament, microwaves, combustion flame, etc. The use of hydrogen facilitate removal of non-diamond deposits, whereas the diamond film is allowed to nucleate on the substrate and grow. As no metal catalysts are used in the synthesis process, the CVD diamond is free from metallic inclusions and therefore it is thermally stable.

1.2.5.1 Free-standing plates

Recent developments in the CVD diamond deposition techniques have enabled fabrication and commercialisation of free-standing diamond plates up to around 1.2 mm thick. These plates are cut to size and shape required by a particular application by laser or EDM. The latter technique, however, may be used exclusively for boron-doped CVD diamond grades that are electrically conductive [73]. To date, the thick-film CVD diamond has mostly been used to complement some PCD products, such as dresser logs [74,75], turning inserts [76], and surgical blades [74].

1.2.5.2 Coated tools

Thus far CVD diamond coated drills and turning inserts have not made significant progress in industrial application mainly due to problematic preparation of the cutting edge and the ease of delamination of the diamond layer [18]. To overcome the latter limitation, special grades of SiC [20], Si_3N_4 and TiC/TiN pre-coated cemented carbide [77] have been utilised as the most suitable substrate materials. The diamond coating has a thickness of typically 20–30 μm and covers only a small working part of the tool [20].

The CVD diamond-coated tools seem to have a great potential for industrial exploitation, especially in machining graphite, metal matrix composites and SiAl alloys [17]. However, further technological improvements are necessary to increase their quality and to decrease the fabrication costs that are still too high, particularly with regard to PCD [18].

References

1. Properties of diamond. De Beers Industrial Diamond Division, Special publication K4000/5/89.
2. Hughes, F.H., The early history of diamond tools. Industrial Diamond Review, 1980, 40(6), 405–407.
3. Hughes, F., Diamond Grinding of Metals. Industrial Diamond Information Bureau, Ascot, 1978, pp. 1–3.
4. Tolansky, S., Early historical uses of diamond tools. In Proceedings of the International Industrial Diamond Conference 'Science and Technology of Industrial Diamonds', edited by Burls, J., Vol. 2, Industrial Diamond Information Bureau, London, 1967, pp. 341–349.
5. Gauthier, E., Diamond lap. U.S. Patent 1,625,463 (April 19, 1927).
6. Jones, W.D., Fundamental Principles of Powder Metallurgy. Edward Arnold Publishers Ltd., London, 1960, p. 807.
7. Brock, F.P., Abrasive implement and method of making same. U.S. Patent 1,537,454 (May 12, 1925).
8. Sanford, B., Abrasive wheel and a method of making the same. U.S. Patent 1,981,970 (November 27, 1934).
9. Lundblad, E., Swedish synthetic diamond scooped the world 37 years ago. Indiaqua, 1990, 55(1), 17–23.
10. Bundy, F.P., Hall, H.T., Strong, H.M., Wentorf, R.H., Man-made diamond. Nature, 1955, 176(4471), 51–55.

11. Bovenkerk, H.P., Bundy, F.P., Hall, H.T., Strong, H.M., Wentorf, R.H., Preparation of diamond. Nature, 1959, 184(4693), 1094–1098.
12. Hall, H.T., Strong, H.M., Wentorf, R.H., Method of making diamonds. U.S. Patent 2,947,610 (August 2, 1960).
13. Jennings, M., ... and the next 50 years? Industrial Diamond Review, 2003, 63(1), 15.
14. Harris, G.I., Long life resin bond wheels. In Ultrahard Materials in Industry - Grinding Metals. De Beers Industrial Diamond Division, 1991, pp. 2–4.
15. Anon., Diamond beads 'Turbo'. Acimm per il Marmo, 1993, 47(5), 149–151.
16. Martin, J., The market: 1940–1990. Industrial Diamond Review, 1990, 50(6), 291–293.
17. Tillmann, W., Trends and market perspectives for diamond tools in the construction industry. International Journal of Refractory Metals & Hard Materials, 2000, 18, 301–306.
18. Clark, I.E., Sen, P.K., Advances in the development of ultrahard cutting tool materials. Industrial Diamond Review, 1998, 58(2), 40–44.
19. Owers, C., Industrial diamond: applications, economics and a view to the future. Industrial Diamond Review, 2000, 60(3), 176–181.
20. Sen, P.K., CVDITE - a new type of cutting tool insert. Industrial Diamond Review, 1992, 52(5), 228–230.
21. Anon., Diamond tools for the new millennium. Marmo Macchine International, 1995, 9, 376–391.
22. Anon., Diamond working well at Intertech 2003. Industrial Diamond Review, 2003, 63(3), 14–15.
23. A review of diamond and CBN sizing & standards. Industrial Diamond Association of America, Inc., Skyland, USA, Publication No. S&S593 5M.
24. Bjerregaard, L., Geels, K., Ottesen, B., Rückert, M. Metalog guide. Struers Tech A/S, Rødovre, 1992.
25. Prekwinkel, H., Single-crystal diamond tools for laminated floors. Industrial Diamond Review, 1997, 57(2), 44–46.
26. Bex, P.A., SYNDIE wire drawing die blanks. In Advances in Ultrahard Materials Application Technology, edited by Daniel, P., Vol. 2, De Beers Industrial Diamond Division, Ascot, 1983, pp. 81–91.
27. Herbert, S.A., Micron diamond - an advancing technology. In Advances in Ultrahard Materials Application Technology, edited by Daniel, P., Vol. 2, De Beers Industrial Diamond Division, Ascot, 1983, pp. 112–125.
28. Ladd, R., Manufactured large single crystal diamond. Paper presented during Intertech 2003, Vancouver, Canada, July 28–August 1, 2003.
29. Sen, P.K., Synthetic diamond dresser logs: serving the future needs of industry. Industrial Diamond Review, 2002, 62(3), 194–202.
30. de Heus, P.R., The applications and properties of MONOCRYSTAL. Industrial Diamond Review, 1997, 57(1), 15–18.
31. Seal, M., The wear of diamond. In Proceedings of the International Industrial Diamond Conference 'Science and Technology of Industrial Diamonds', edited by Burls, J., Vol. 1, Industrial Diamond Information Bureau, London, 1967, pp. 145–159.
32. Stalder, H., Applications of MONOCRYSTAL whole stones. Industrial Diamond Review, 2000, 60(1), 48–49.
33. O'Brien, P., Monodie makes more wire. Industrial Diamond Review, 1992, 52(2), 321–322.
34. Anon., Syndie & Monodie - the largest die blank range in the world. Industrial Diamond Review, 1991, 51(4), 168–172.
35. Smith, N.R., User's Guide to Industrial Diamonds. Hutchinson Benham, London, 1974, pp. 43–44.
36. Danger, K.-H., Küllmer, M., Diamond tools used in dentistry. Industrial Diamond Review, 1999, 59(1), 73–74.
37. Bakoń, A., Szymański, A., Practical Uses of Diamond. Ellis Horwood & Polish Scientific Publishers, Warsaw, 1993, pp. 124–130.
38. Kopp, O., Method for manufacturing metal wires partially covered with abrasive particles for cutting tools. U.S. Patent 4,852,998 (August 1, 1989).
39. Li, Y., Li, G., Jiang, H., He, Y., New type of matrix material for the manufacture of electroplated diamond tools. Industrial Diamond Review, 2002, 62(4), 259–262.

40. Sofer, Y., Yarnitzky, Y., Dirnfeld, S.F., Evaluation and uses of composite Ni-Co matrix coatings with diamonds on steel applied by electrodeposition. Surface and Coating Technology, 1990, 42, 227–236.
41. Musu-Colman, M., Fecioru, M., Baluta, G., Arnici, N., Surface processing of ultrahard materials used for embedding in resin or metallic matrices. In Proceedings of Powder Metallurgy World Congress & Exhibition, Granada, Spain, October 18–22, 1998, Vol. 4, pp. 234–239.
42. Cooley, B.A., Juchem, H.O., Vitrified bonds – no longer a synonym for conventional abrasive tools. In Proceedings of Superabrasives'85, Chicago, U.S.A., April 22–25, 1985, pp. 2.15–2.56.
43. Przyklenk, K., Diamond impregnated tools - uses and production. Industrial Diamond Review, 1993, 53(4), 192–195.
44. Premadia – premium diamond abrasives for vitrified bonds. De Beers Industrial Diamond Division. Technical data leaflet NC3000493.
45. Bailey, M.W., Juchem, H.O., Selection and use of Premadia. Industrial Diamond Review, 1994, 54(1), 8–11.
46. Man-made diamond products for metal, plated, and vitreous bonds. GE Superabrasives. Product information leaflet GES 91-968.
47. Antonini F., Vitrified bonds. Diamante Applicazioni & Tecnologia, 1994, 1(2), 16–17.
48. Man-made diamond products for resin bonds. GE Superabrasives. Product information leaflet GES 91-969.
49. Premadia - premium diamond abrasives for resin bonds. De Beers Industrial Diamond Division. Product information leaflet NC3000493.
50. RVG diamond products. Real value grinding. GE Superabrasives, 2002. Product information brochure GES 1320 E.
51. Jakobuss, M., The dynamics of diamond retention in grinding wheel systems. Paper presented during Intertech 2000, Vancouver, Canada, July 17–21, 2000.
52. Jennings, M., Special tooling for specialist tools. Industrial Diamond Review, 1998, 59(1), 6–7.
53. Bailey, M.W., Garrard, R., Juchem, H.O., Characteristics of diamond and their effect on grinding behaviour. Industrial Diamond Review, 1999, 59(1), 10–19.
54. von Mackensen, V., Longerich, W., Dennis, P., Preising, D., Fine grinding with diamond and cBN. Industrial Diamond Review, 1997, 57(2), 40–43.
55. Schwan, G., Zimmerer, R., Economic polishing of granite, Industrial Diamond Review, 1998, 58(1), 4–5.
56. Cook, M.W., Diamond machining of MMC engineering components. Industrial Diamond Review, 1998, 58(1), 15–18.
57. Clark, I.E., PCD wood tools – a new design concept. Industrial Diamond Review, 1993, 53(2), 73–76.
58. Feenstra, R., Status of polycrystalline-diamond-compact bits: Part 1 – Development. Journal of Petroleum Technology, 1988, 40(6), 675–684.
59. Walmsley, J.C., Lang, A.R., TEM study of SYNDAX3, compared with SYNDITE and AMBORITE. In Advances in Ultrahard Materials Application Technology, edited by Barrett, C., Vol. 4, De Beers Industrial Diamond Division, Ascot, 1988, pp. 61–75.
60. Compax diamond die blanks. GE Superabrasives, 2000. Product information brochure GES 1248.
61. Clark, I.E., Polycrystalline diamond drill bits in mining applications. In Advances in Ultrahard Materials Application Technology, edited by Barret, C., Vol. 4, De Beers Industrial Diamond Division, Ascot, 1988, pp. 17–35.
62. Tomlinson, P.N., Clark, I.E., SYNDAX3 pins – new concepts in PCD drilling. Industrial Diamond Review, 1992, 52(3), 109–114.
63. Ersoy, A., Waller, M.D., Drilling detritus and operating parameters of thermally stable PCD core bits. International Journal of Rock Mechanics and Mining Sciences, 1997, 34(7), 1109–1123.
64. Fleming, M.A., Sen, P.K., The role of wire drawing dies and the factors influencing die life. Industrial Diamond Review, 2000, 60(1), 29–35.
65. Introduction to De Beers PCD and PCBN cutting tool materials. De Beers Industrial Diamond Division. Product information brochure No. 1.2.1.
66. Compax diamond tool blanks for machining nonferrous and nonmetallic materials. GE Superabrasives, 2001. Product information brochure GES 1309 E.

67. Introduction to De Beers PCD drilling materials. De Beers Industrial Diamond Division. Product information brochure No. 1.2.2.
68. Stratapax drill blanks & Geoset drill diamond. GE Superabrasives, 2002. Product information brochure GES 1327 E.
69. Spur, G., Appel, S., Wire EDM cutting of PCD. Industrial Diamond Review, 1997, 57(4), 124–130.
70. Anon., New laser dicing machine for PCD fabrication. Industrial Diamond Review, 2003, 63(1), 38–39.
71. Clark, I.E., Mackenzie, K.A., Tomlinson, P.N., Field results in drilling & mining with SYNDAX3. Industrial Diamond Review, 1992, 52(4), 186–191.
72. Derjaguin, B.V., Fedoseev, D.V., Low pressure diamond growth. Industrial Diamond Review, 1990, 50(3), 155–160.
73. Olsen, R.H., Dewes, R.C., Aspinwall, D.K., Collins, J., Cook, M., Electrical discharge machining of conductive CVD diamond and PCD. Paper presented during Intertech 2003, Vancouver, Canada, July 28–August 1, 2003.
74. Collins, J.L., New CVD diamond products. Industrial Diamond Review, 1999, 59(3), 212–213.
75. Pricken, W., Dressing of vitrified bond wheels with CVDRESS and MONODRESS. Industrial Diamond Review, 1999, 59(3), 225–231.
76. Brookes, K., MIM raises the stakes in hardmetal manufacturing. Metal Powder Report, 2004, No. 1, 23–27.
77. Faure, C., Hänni, W., Julia Schmutz, C., Gervanoni, M., Diamond-coated tools. Diamond and Related Materials, 1999, 8, 830–833.

CHAPTER 2

Machining with Diamonds – Theoretical Model

It is of critical importance to the diamond tool design and fabrication that the toolmaker fully understands the complex interactions occurring between the tool and machine, the workpiece and the application conditions. Figures revealed recently indicate that the PM products account for around 62% of all diamond-impregnated tools [1]. Interestingly, production of sawblades and drills for stone cutting and construction applications has been identified as the biggest segment accounting for around 61% of the overall consumption of industrial diamonds in Europe being the largest market worldwide [2]. Therefore, for obvious reasons, it seems justified to provide in this chapter analytical descriptions of machining of a ceramic-type workpiece with diamonds in various sawing and drilling operations. As both sawing and drilling with diamond-impregnated tools are generally considered as grinding processes with geometrically indeterminate cutting edges, the proposed models may readily be fitted for most grinding operations as well.

2.1 Circular Sawing

Sawing brittle materials, such as stone, concrete, ceramics, etc., by means of a diamond-impregnated saw consists in wearing away its mineral constituents by passing rigid grits over the machined surface. The diamond crystals, acting as cutting edges, are firmly held in a matrix, which wears progressively exposing fresh particles while those protruding sufficiently to cut are subjected to mechanical degradation to be finally dislodged.

In order to attain the economically best sawing conditions, an ideal balance between the tool life and cutting rate has to be achieved. The harder the workpiece to be cut, the stronger the diamond type to be selected is a general rule while the matrix has to erode at a rate compatible with the diamond breakdown. An incorrect choice of the matrix and/or the diamond type, size and concentration yields a tool that wears away excessively or refuses to cut altogether.

Regardless of the choice of diamond and matrix, there are other factors, which have a strong effect on the sawblade performance. The most important are:

- segment, or cutting rim, manufacturing method and parameters [3–11],
- workpiece properties [11–19],
- sawing conditions [13,17,20–26],
- cooling efficiency [27–29],
- quality of the segment-to-metal core joint [30–34],
- metal core design and tensioning [22,24,35–41],
- machine condition and operator's skills [13,17,42].

Obviously, all the above-mentioned items are closely interrelated and therefore each of them may prove critical and must not be neglected.

With circular sawing, the blade rotates in a constant direction at high peripheral speeds of typically 25–65 m/s. This leads to the development of a tail of matrix behind each individual diamond particle, as shown in Fig. 2.1, which acts as a support during cutting.

To provide enough space for chip removal, parameters characterising the sawing operation have to be considered in conjunction with the composition of the cutting rim so not to thicken the slurry excessively, thus avoiding harsh wear conditions for the matrix. The direction of sawing, whether upwards or downwards, may also affect the tool behaviour and justify a modification to the sawblade specification.

Its effect is two-fold. First, the relative orientation of the resultant cutting force and the support frame is different in down- and up-cutting as shown in Fig. 2.2.

In either mode unstable sawing conditions appear when the resultant force deviates too much from the direction of maximum rigidity, thus generating harmful vibrations in the system [43–45]. Assuming that the spindle support in Fig. 2.2 is positioned

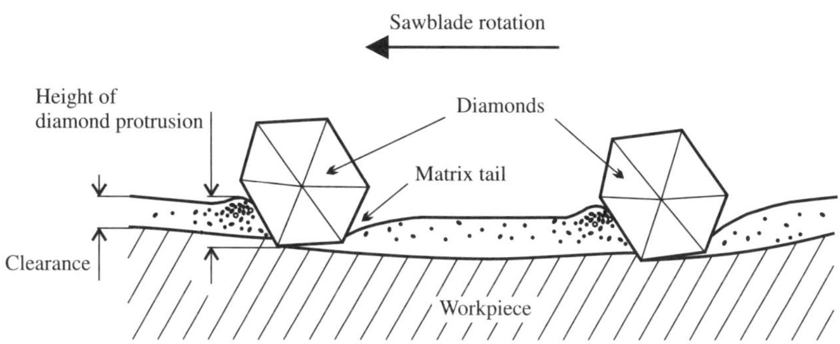

Figure 2.1 Schematic representation of the cutting zone in circular sawing.

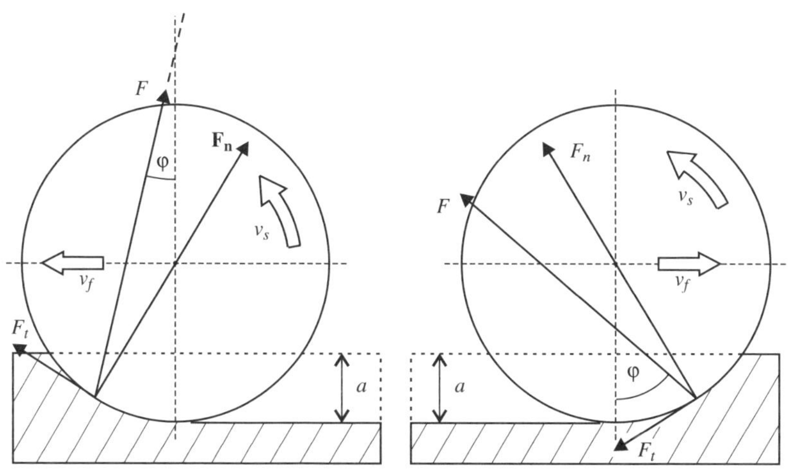

Figure 2.2 Forces acting on a sawblade in down-cutting (left) and up-cutting (right) modes.

vertically, an increasing ratio of the tangential to normal force (F_t/F_n) results in lower stability in the up-cutting mode while in down-cutting an opposite effect is observed.

Second, the diamond loading conditions change dramatically when reversing the sawblade rotation as illustrated in Fig. 2.3.

In down-cutting, diamond particles penetrate into the workpiece to full depth while coming in contact with it. Thereafter as the diamonds track across the abraded surface layer, they emerge progressively and finally lose contact with the workpiece. In contrast to down-cutting, in the up-cutting mode diamonds gradually increase the depth of penetration to achieve the maximum values while leaving the kerf. Consequently, the diamond breakdown is facilitated by downward rotation of the blade that, by its nature, creates additional, pulsing impact forces acting on crystals, which enter the cutting zone.

It has been proved experimentally that the wear pattern over the circumference of a sawblade depends on the cutting mode as well [46]. In up-cutting all the segments are subjected to wear at fairly the same rate, whereas in down-cutting the wear rate of an individual segment is variable. Sites prone to accelerated wear tend to move around the sawblade periphery in the direction opposite to that of its rotation. Such a 'spiral' wear process is attributed to the aforementioned impact pulse loading of the cutting diamond grit causing increased removal of material in the kerf of an order which prevents the subsequent diamonds from engaging [46].

Assuming that a workpiece is cut by a thin sawblade that contains a single raw of equally spaced diamond cutting edges of the same width as the steel centre and located at the same height, then the chip removed by an individual diamond particle attains the maximum thickness h_{max} [11,47],

$$h_{\max} \approx \frac{Z_A}{v_s C w} \sqrt{\frac{1}{aD} - \frac{1}{D^2}} \tag{2.1}$$

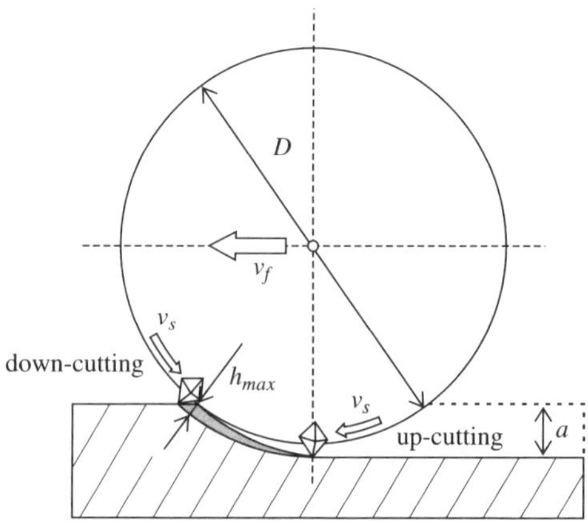

Figure 2.3 Kinematic diagram of diamond crystals that come in contact with workpiece in down-cutting (left diamond) and up-cutting (right diamond) modes.

where Z_A is the cutting rate (being a product of feed rate v_f and depth of cut a), v_s the blade peripheral speed, C the surface concentration of cutting edges, w the width of cutting edge, a the depth of cut and D the blade diameter.

It is noteworthy that by putting together a number of such thin sawblades and rotating each of them on the spindle at a random angle it is possible to produce a simplistic model of a continuous rim sawblade with a quasi-random distribution of cutting edges on its periphery as demonstrated in Fig. 2.4.

Although it is difficult to approximate the actual values of C and w, relation (2.1) has an important practical meaning. It quantifies the effect of the combination of diamond related parameters, sawing conditions and tool size on the maximum chip thickness, thus aiding the toolmaker in optimising the segment composition. Although in the field practice, the severity of segment wear conditions is determined by many factors and independent judgement of a particular situation must always be exercised, the maximum chip thickness appears to be a reliable preliminary indicator [11].

Another important parameter is the local, cumulative chip thickness H, defined as the instantaneous thickness of the overall chip accumulated at a certain position between the cutting rim and the workpiece [11,47]. The local value of H depends on the direction of sawing, angular spacing of the cutting edges and their position. Assuming that the cuttings, carried by the coolant, are moving in the same direction as the blade at a constant rate of $v_s/2$, the cumulative chip thickness distribution throughout the cutting zone is different for down- and up-cutting conditions, as shown in Fig. 2.5 [11,47].

In down-cutting, large amounts of debris are generated by diamonds entering the zone, thus leading to rapid thickening of the slurry which promotes harsher wear conditions for the matrix. In up-cutting, the slurry thickening process is slower since diamonds attain the maximum material removal rate when leaving the zone. Importantly, for a continuous rim blade the mean cumulative chip thickness at the

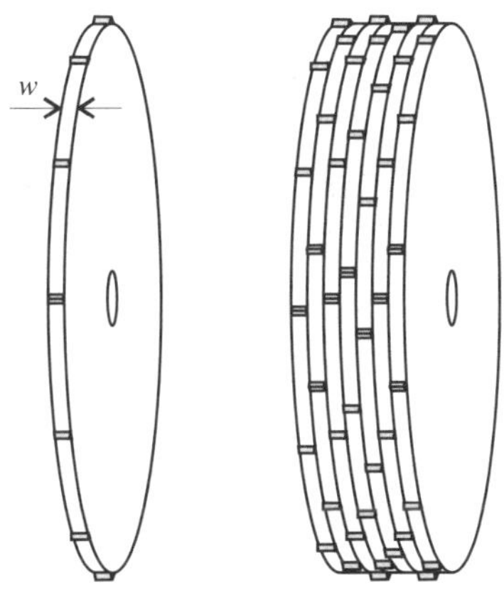

Figure 2.4 Continuous rim sawblade used to model sawing operations.

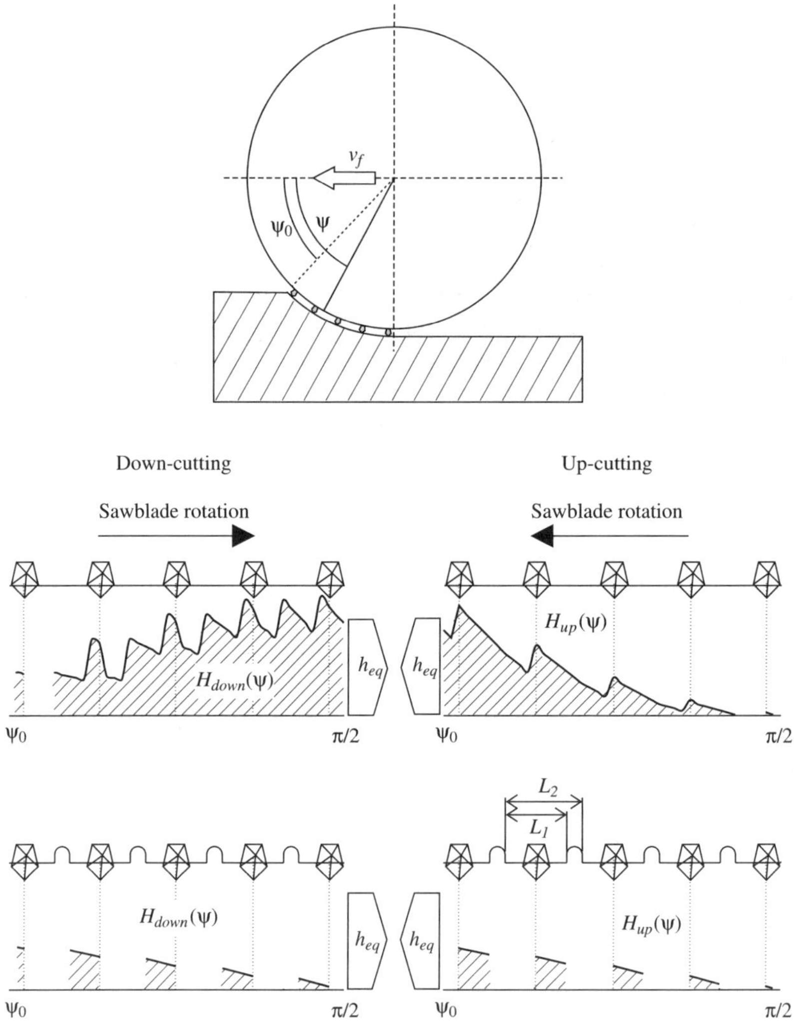

Figure 2.5 Numerically generated cumulative chip distribution diagrams in down-cutting (left) and up-cutting (right) for a continuous rim (centre) and segmental (bottom) blade.

outlet of the cutting zone does not depend on the cutting mode and can be derived from the following relation [11,47]:

$$\overline{H}_{\mathrm{down}}(\pi/2) = \overline{H}_{\mathrm{up}}(\psi_0) \cong \frac{Z_{\mathrm{A}}}{v_{\mathrm{s}}} = h_{\mathrm{eq}} \tag{2.2}$$

where h_{eq} is commonly referred to as the equivalent chip thickness.

The situation changes dramatically when the workpiece is cut by a segmental sawblade. The debris generated by each working diamond travel along the trailing part of the segment to be finally expelled into a slot behind it. In this case, expression (2.2) is no longer valid since certain volume of the cuttings, which depends on the blade design, is trapped in the slots. The mean cumulative chip thickness depends on the length of the trailing section of the segment which indicates the advantage of shorter segments, with the same rim-partition ratio (L_1/L_2), when excessive matrix wear is observed.

It has to be emphasised that slots incorporated into the cutting rim of a sawblade prevent excessive accumulation of debris in the cutting zone. However, a reliable quantification of H for this particular case is problematic because unknown amount of debris travelling with a slot is sucked back into the cutting zone of the incoming segment.

2.2 Frame Sawing

In frame sawing, which has long been utilised on a broader scale in sawing stone blocks, the blade is subjected to a reciprocal sawing action at a slow sinusoidal speed with a maximum of around 2 m/s. The reciprocating movement of the blade slows down the removal of stone detritus from the cut. This creates severe wear conditions for the matrix and facilitates diamond pullout since there is no build-up of matrix tail, and forces act on diamond grits in alternate directions. Schematic illustration of the cutting zone characteristic of frame sawing is presented in Fig. 2.6.

By analogy to circular sawing, in the frame sawing model it is assumed that a stone block is cut by a thin, linear sawblade that contains a single raw of equally spaced cutting edges of the same width as the blade and located at the same height. From Fig. 2.7 it is evident, that the instantaneous horizontal speed of the sawblade depends on the flywheel position.

Under such conditions, the maximum chip thickness is attained while altering the movement direction and may be approximated as [11,47],

$$h_{\max} \approx \frac{v_f}{n_s} \arccos\left(\frac{l_s - 2L_3}{l_s}\right) \tag{2.3}$$

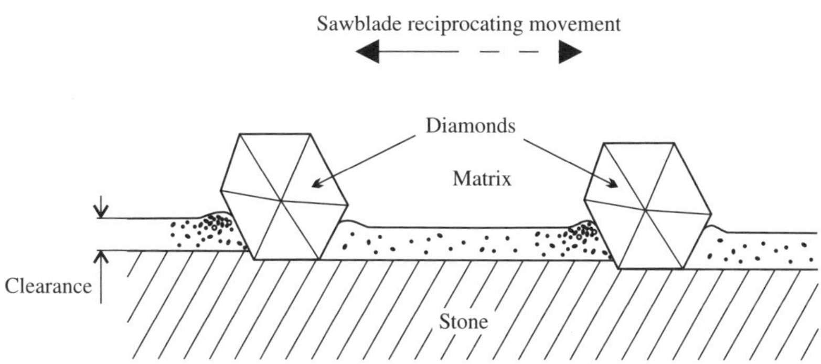

Figure 2.6 Schematic representation of the cutting zone in frame sawing.

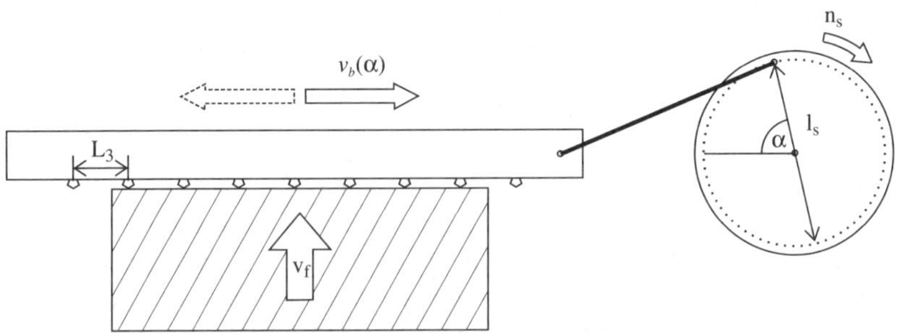

Figure 2.7 Kinematic diagram of frame sawing operation.

where n_s is the flywheel rotational speed (number of rotations per minute), l_s the length of stroke and L_3 the spacing of cutting edges.

By contrast to circular sawing, the frame saw acts at a slow horizontal speed and consequently the impact on the diamond is low. Therefore, the matrix resistance to wear takes priority over the diamond strength.

Obviously application of high feed rates in combination with a small number of insufficiently protruding cutting edges results in high penetration depths and hence short tool life. From relation (2.3), however, it becomes evident that harsh wear conditions are also created on so-called 'slow frames' [48], which are characterised by low flywheel rotational speed and small length of stroke.

2.3 Wire Sawing

A diamond wire has become a standard stone quarrying tool. Due to its adaptability to suit most sawing tasks, improved sawn-surface texture, reduced noise and vibration; it is also being increasingly used in a variety of stone and concrete sawing operations [49–52]. In contrast to any other type of metal-bonded tools, in a wire saw the diamond-impregnated beads are attached at regular intervals to a flexible support which is composed of many high strength stainless-steel strands twisted together to form a rope.

The cutting action consists in pulling a properly pre-tensioned wire saw across the workpiece as schematically presented in Fig. 2.8.

It is essential for the tool performance that the diamond beads are firmly fixed in place on the rope and wear in a uniform manner over the whole working surface. Therefore the wire saw should rotate in the kerf, but the beads must not rotate around the steel rope holder [53]. In industrial practice a correct selection of the sawing method usually yields satisfactory results [53,54], but pre-twisting the wire around its centre axis before a continuous loop is assembled [54,55] and a proper diamond bead design [56] may also aid in its uniform wear.

Contrary to circular and frame saws, the diamond wire has no clearly defined shape and is considered to be completely flexible. Therefore the simple model used to

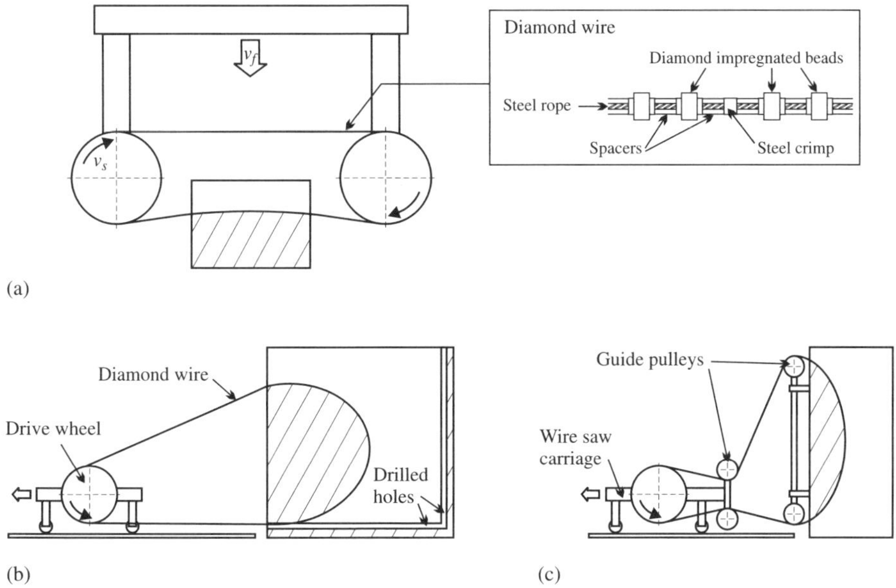

Figure 2.8 Layouts of typical wire sawing operations on a stationary machine (a) and quarry-type machines (b) and (c).

derive equations (2.1) and (2.3) cannot be applied to wire sawing operations, except in the case shown in Fig. 2.8a, for which the maximum chip thickness may be estimated from the equation

$$h_{\mathrm{max}} = \frac{v_{\mathrm{f}} L_3}{v_{\mathrm{s}}} \tag{2.4}$$

where v_{s} is the wire linear speed. It should be emphasised, however, that equation (2.4) holds when the wire has already been fully engaged and its deflection is small and sustainable at a near-constant level.

In the quarry-type machines, presented in Figs. 2.8b and 2.8c, both the length and shape of the cutting zone vary markedly with time and hence no simple algorithm exists whereby h_{max} could be estimated. As a further complication a contribution from the wire driving system to some scatter of forces acting on the working part of the diamond wire is also expected.

In the simplistic model of the diamond wire cutting zone, shown in Fig. 2.9, it is assumed that the forces acting on a single cutting point are in equilibrium.

For small values of α the normal diamond loading force F_{l} can be approximated as follows:

$$F_{\mathrm{l}} \cong 2F_{\mathrm{t}} \sin(\alpha) - F_{\mathrm{c}}$$

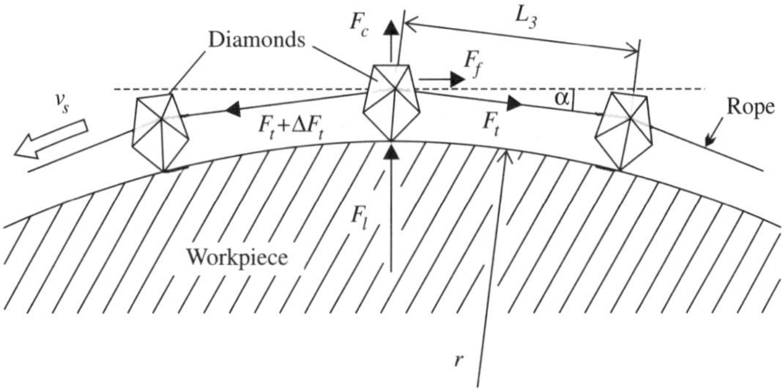

Figure 2.9 Forces acting on a diamond wire.

where F_t is the rope tensioning force and F_c the centrifugal force.

Substituting

$$F_c = \frac{m_L L_3 v_s^2}{r}, \quad \sin(\alpha) = \frac{L_3}{2r}$$

where m_L is the mass per unit length of the wire saw and r the local radius of curvature of the cutting zone.

The loading force may finally be evaluated as follows:

$$F_l \cong (F_t - m_L v_s^2)\frac{L_3}{r} \quad (2.5)$$

The loading force has a direct bearing on the diamond penetration depth and, as it may be inferred from the existing experimental data [57], there is a virtually perfect linear relationship between these two parameters. Therefore equation (2.5) implies that, when the operational variables are fixed, the theoretical local chip thickness depends on the wire saw curvature.

Initially, as illustrated in Fig. 2.10, the wire bends sharply at the corners (A), but has a tendency to adopt more regular shape (B) as sawing proceeds. It is quite common that the cut is far from completion when the wire saw carriage approaches the end of the track. Therefore the wire saw has to be shortened and, immediately after sawing has been restarted, the rope is forced away from its rectilinear movement when it enters and leaves the cutting zone (B'). Thereafter the wire saw loop smoothes round (C) and its curvature increases (D).

As it is evident from the above example, the configuration of the cutting zone is complex and alters with time, and hence there is no simple method whereby the chip thickness distribution throughout the cutting zone could be assessed. Attempts have been made to gain more detailed insight into the wire sawing mechanisms by computer simulation [57,58]. Even though the preliminary results are promising,

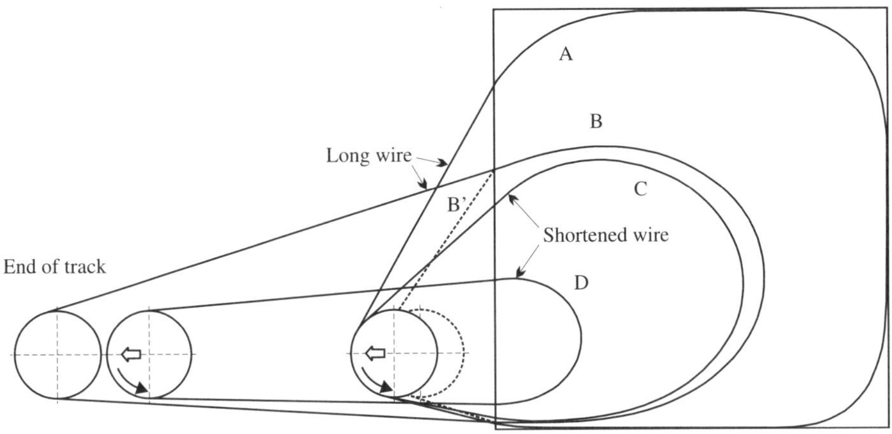

Figure 2.10 Progression of sawing with a diamond wire.

numerical modelling still lacks versatility and, as yet, has not become instrumental in optimising the bead composition and wire saw design.

2.4 Core Drilling

In core drilling, the bit rotates in a constant direction at peripheral speeds of between 1 and 10 m/s [59] and with its active part remaining in a quasi-continuous contact with the workpiece material. This leads to the development of matrix tails behind the exposed diamonds, as in the case of circular sawing (see Fig. 2.1).

Figure 2.11 implies that the kinematic conditions of chip formation in core drilling are relatively simple as opposed to the previously discussed diamond sawing techniques.

Assuming that the drilling crown contains a single raw of equally spaced diamond cutting edges of the same width as the core barrel and located at the same height, then the chip removed by an individual diamond particle has a constant thickness,

$$h = h_{\mathrm{max}} = \frac{v_{\mathrm{f}}}{v_{\mathrm{s}} C w} \tag{2.6}$$

The closed-ring geometry of the cutting zone imposes severe restrictions on evacuation of drilling swarf from the kerf. In order to prevent excessive tool wear by abrasion it is essential that a cooling fluid flows through the centre of the coring drill barrel in sufficient amounts to wash the cuttings away. In most cases, water is pumped into the narrow channel between the core barrel and the drilled material. To ensure easy fluid circulation the drill bit crown has to possess either segmental or quasi-continuous rim design with special waterways incorporated into it in the latter case.

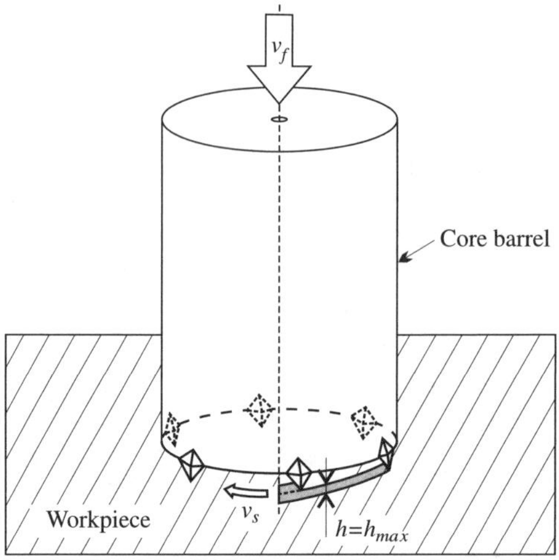

Figure 2.11 Kinematic diagram of core drilling operation.

2.5 Concluding Remarks

In actual situations, the cutting edges are never located at the same height but are distributed at random. Therefore, merely a small proportion of diamonds retained on the tool surface interact with the workpiece being subjected to particle breakdown cycles wherein the height of protrusion and geometry of each cutting edge alter progressively [60].

Furthermore, the chip formation process is influenced in a complex manner by additional factors such as

- instantaneous topography of the machined surface,
- workpiece properties (mineralogical composition, strength, hardness, grain size, etc.),
- magnitude of forces acting between the tool and the workpiece,
- stress distribution in the workpiece,
- temperature generated and its distribution throughout the cutting zone,

as well as many other system dependent variables and incidental events.

The existing theories of sawing of natural stone and ceramic materials discriminate between primary and secondary chip formation processes [61,62]. The former operates in front of each cutting diamond where the debris is produced by an action of tensile and compressive stresses generated alternately in the workpiece by a moving diamond. The primary chip contributes significantly to formation of a front crater and chip removal grooves behind each working diamond as illustrated in Fig. 2.12.

Concurrently, the workpiece material is deformed by compressive stresses generated beneath the cutting diamonds. As soon as the load has been removed an elastic

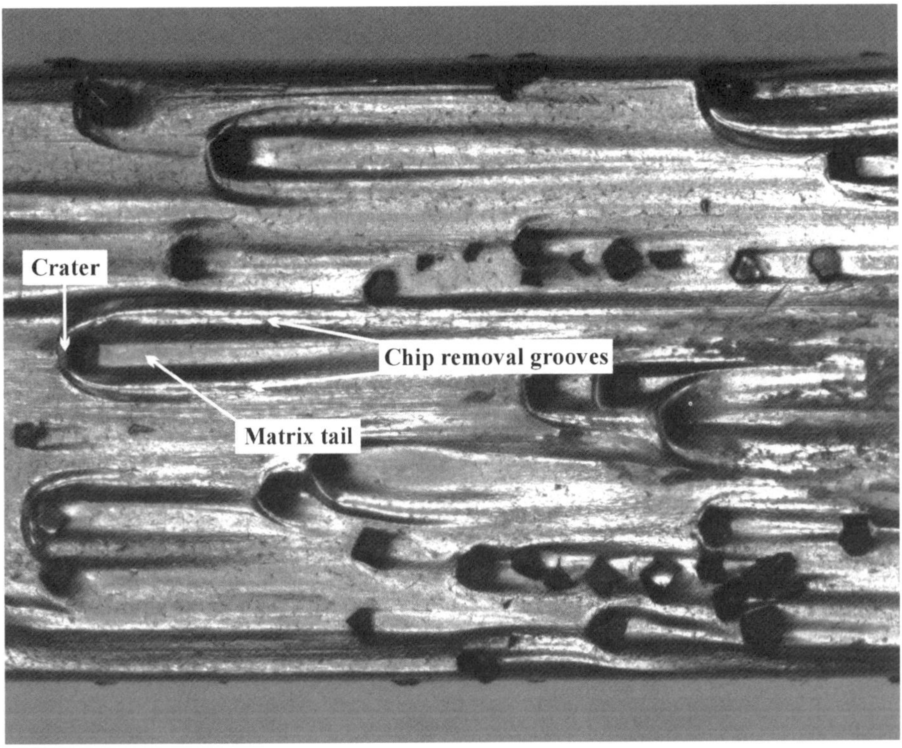

Figure 2.12 Surface topography of a circular sawblade segment.

reversion leads to critical tensile stresses, which cause brittle fracture whereby secondary chip is produced behind the diamond.

It has been well documented that when scratching brittle ceramics the material removal consists in chip formation by plastic flow and brittle fracture mechanisms [63–68]. When the diamond penetration depth is sufficient to cause cracks to appear, a chip is created due to brittle fracture of the material. As presented schematically in Fig. 2.13, there is a zone of irreversible plastic deformation below the groove made by tangential movement of an abrasive grain along the abraded surface and two main crack systems, radial and lateral, derived from that zone.

The radial cracks, which are considered to cause strength degradation, are created by the action of the wedge-shaped abrasive when high normal load is applied and they may continue to propagate as the load is removed due to the residual tensile stresses acting on the tip of the crack. The lateral cracks are initiated when the load is removed and may propagate to carry on with the residual stress relaxation process. Material removal by fracture is generally associated with lateral cracking provided that the force on the wedge-shaped indentor is above a certain critical value which depends on the mechanical properties of the workpiece material.

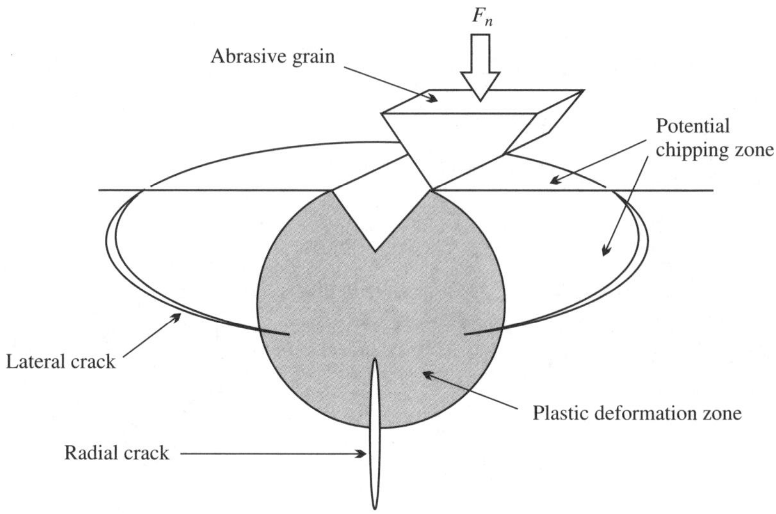

Figure. 2.13 Schematic representation of the plastic deformation zone, radial and lateral cracks produced by scratching brittle material with a wedge-shaped abrasive [64–67].

It has been found experimentally that the specific material removal energy, defined as

$$E_{sg} = \frac{F_t}{wh_{eq}} \tag{2.7}$$

strongly depends on the equivalent chip thickness [66]. For low h_{eq} values ductile flow is the predominant mechanism and no crack formation is observed. This results in intense heat production by friction and higher tangential cutting force due to wear-flatting of diamond particles. Conversely, for high h_{eq} values there is an intense formation of both radial and lateral cracks, which are responsible for the reduction in workpiece strength and easier material removal respectively. Therefore, as the material is already fragile it may partially be removed by a raking action of the tool which lowers the tangential force.

The description of material removal process based on the equivalent chip thickness h_{eq} has, however, a serious shortcoming as it takes no account of factors that affect the cutting forces.

First, the brittle fracture zone does not appear until a certain critical chip thickness has been attained [68] and the crack formation process is apparently affected by the magnitude of the impact forces acting on the workpiece. It is therefore reasonable to assume that F_t, and consequently E_{sg}, depends on the maximum chip thickness, i.e. on the kinematics of the cutting process.

Industrial practice has given evidence of this [69]. As shown in Fig. 2.14, for constant cutting rate and blade peripheral speed, which renders h_{eq} constant as well, the power consumption in circular sawing of granite, which is a product of F_t and v_s, increases when deeper cuts are made, i.e. when h_{max} is being reduced.

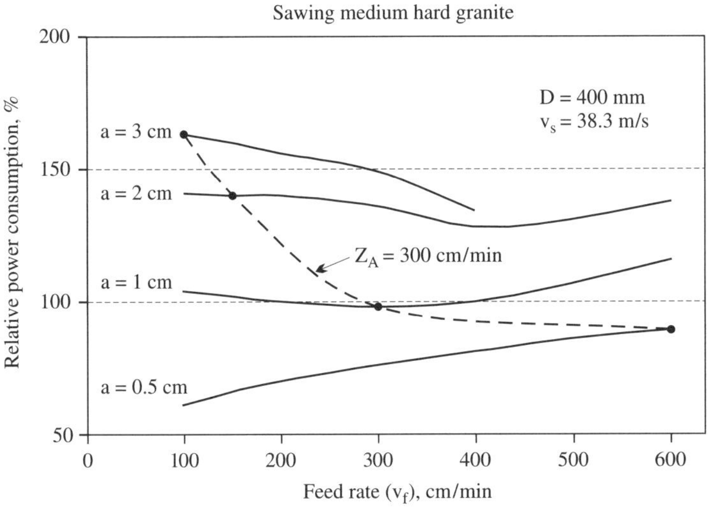

Figure 2.14 The dependence of power consumption on feed rate and depth of cut (adapted from Ref. [69]).

Table 2.1 Results of sawing tests performed on class IV granite (adapted from Ref. [26]).

Cutting rate Z_A (cm^2/min)	Depth of cut a (mm)	Length of contact l_g (cm)	F_n/l_g (N/cm)	F_t/l_g (N/cm)	Power drawn (kW)	h_{max}[1]
150	7.5	3.9	120	11.1	1.27	1.00
300	10	4.5	73	11.6	1.53	1.72
300	7.5	3.9	56	11.4	1.29	2.00
450	15	5.5	50	12.4	2.14	2.08

[1]Results are relative to Z_A=150 cm^2/min (h_{max}=1.00).

Second, detailed analysis of forces acting on a circular sawblade indicates that the normal to tangential force ratio strongly depends on the process parameters [26]. As documented in Table 2.1, the tangential force per unit length of contact remains virtually unaffected by the sawing conditions whereas its normal counterpart varies to a large extent.

Interestingly, the normal force per unit length of contact ideally correlates with the maximum chip thickness. The perfect inverse relation between the normal load per unit length of contact and maximum chip thickness implies dulling of the cutting points at low diamond penetration depths.

References

1. Przyklenk, K., Diamond impregnated tools – uses and production. Industrial Diamond Review, 1993, 53(4), 192–195.
2. Tillmann, W., Trends and market perspectives for diamond tools in the construction industry. International Journal of Refractory Metals & Hard Materials, 2000, 18, 301–306.

3. Konstanty, J., The materials science of stone sawing. Industrial Diamond Review, 1991, 51(1), 27–31.
4. Konstanty, J., Bunsch, A., Hot pressing of cobalt powders. Powder Metallurgy, 1991, 34(3), 195–198.
5. Konstanty, J., Bunsch, A., Cias, A., Factors affecting hardness and ductility of hot-pressed cobalt powders. Powder Metallurgy International, 1991, 23(6), 354–356.
6. De Chalus, P.A., Konstanty, J., Diamond tooling. Stone cutting. Cobalt News, 1996, (4), 12–17.
7. Konstanty, J., Hot compacting of extrafine cobalt. In Advances in Powder Metallurgy & Particulate Materials – 1996. Publ. Metal Powder Industries Federation, Princeton, NJ, 1996, Vol. 2, Part 5, pp. 3–7.
8. Romanski, A., Frydrych, H., Konstanty, J., Mechanical properties of cobalt-base materials for diamond impregnated tools. In Proceedings of International Workshop on Diamond Tool Production, Turin, Italy, November 8–10, 1999, pp. 191–196.
9. Konstanty, J., Diamond bonding and matrix wear mechanisms involved in circular sawing of stone. Industrial Diamond Review, 2000, 60(1), 55–65.
10. Konstanty, J., Romanski, A., Frydrych, H., Effect of mechanical properties of the matrix on its diamond retention capacity and wear characteristics during sawing hard stone by means of diamond impregnated tools. In Proceedings of Powder Metallurgy World Congress, Kyoto, Japan, November 12–16, 2000, Part 2, pp. 1629–1632.
11. Konstanty, J., Cobalt as a Matrix in Diamond Impregnated Tools for Stone Sawing Applications, AGH Uczelniane Wydawnictwa Naukowo-Dydaktyczne, 2nd Edition, Krakow, 2003.
12. Luce, E.C., Newell, C.R., Sheehy, J., An abrasive scale for granite and its influence on diamond blade evaluation. Diamonds and Stone. Paper presented at the Scandinavian Stone Conference, Gothenburg. Diamond Information L10, De Beers Industrial Diamond Division, October 2–3, 1967, pp. 14–16.
13. Anon., Machining stone with diamond tools. Diamond Information L15, De Beers Industrial Diamond Division.
14. Burgess, R.R., Birle, H.D., Circular sawing granite with diamond sawblades. Paper presented during the Industrial Diamond Association of Japan 30th Anniversary Meeting and Seminar, Tokyo, Japan, May 16, 1978.
15. Wright, D.N., Cassapi, V.B., Factors influencing stone sawability. Industrial Diamond Review, 1985, 45(2), 84–87.
16. Birle, J.D., Ratterman, E., An approximate ranking of the sawability of hard building stones based on laboratory tests. GE Superabrasives, 1988. Booklet GES 71–553.
17. Jennings, M., Wright, D., Guidelines for sawing stone. Industrial Diamond Review, 1989, 49(2), 70–75.
18. Cardu, M., Frisa, M. A., Mancini, R., Marini, P., Investigation research to define a standard laboratory test aimed to foresee the cuttability of ornamental stones by diamond tools. Marmo Macchine, 1996, 26(5), 206–222.
19. Wright, D.N., Tagg, W.R.J., The development of a rock classification system for use with diamond tools. Industrial Diamond Review, 1998, 58(4), 113–120.
20. Bailey, M.W., Bullen G.J., Diamond sawing in the stone and construction industries. De Beers Industrial Diamond Division internal paper.
21 Rigvall, B., The choice of cutting parameters in sawing stone with diamond circular saws. Diamonds and Stone. Paper presented at the Scandinavian Stone Conference, Gothenburg. Diamond Information L10, De Beers Industrial Diamond Division, October 2–3, 1967, pp. 10–13.
22. Collin, W.D., Some aspects of blade and machine stability when sawing concrete. Drilling and sawing concrete with diamond tools. Diamond Information L34, De Beers Industrial Diamond Division, pp. 6–12.
23. Palovchik, S.T., Sawing reinforced concrete with diamond blades. Paper presented at Concrete Sawing and Drilling Association 1975 Annual Convention, New Orleans, Louisiana, March 13–15, 1975.
24. Bailey, M.W., Wright, D.N., SDA85 and SDA100 in circular saws – the effect of cutting parameters. In Proceedings of De Beers Düsseldorf Conference'79, Düsseldorf, Germany, May 22–23, 1979, pp. 2.15.1–2.15.12.
25. Wöll, E., Cutting asphalt with diamonds. Diamonds in Industry – Roads and Runways, De Beers Industrial Diamond Division, 1984, 44–51.

26. Webb, S.W., Jackson, W.E., Analysis of blade forces and wear in diamond stone cutting. Journal of Manufacturing Science and Engineering, 1998, 120, 84–92.
27. Miller, H.C., Where does the water go? In Proceedings of the 1st International Congress on Diamonds in Industry, Paris, France, May 28–June 2, 1962, pp. 217–225.
28. Tönshoff, H.K., Schulze, R., The effect of the coolant in the sawing of hard stone. Industrial Diamond Review, 1980, 40(4), 252–257.
29. Wang, C.Y., Wei, X., Tang, Z.L., Pan, Z.C., The role of coolant in granite sawing. Industrial Diamond Review, 1995, 55(4), 156–160.
30. Clauser, G., Valle, A., Laser welding explained. Diamonds in Industry – Building and Construction. De Beers Industrial Diamond Division, 50–51.
31. Weber, G., Laser welding of diamond tools. Industrial Diamond Review, 1991, 51(3), 126–128.
32. Anon., The brazing book. Handy & Harman, New York, USA, 1991.
33. Weber, G., Burckhardt, S., New laser welding machine for diamond sawblade manufacture. Industrial Diamond Review, 1999, 59(2), 88–89.
34. Ugues, D., Grande, M.A., Rosso, M., Vaile, A., Advances in laser welding for cutting diamond tools production. In Proceedings of PM2001 Congress and Exhibition, Nice, France, October 22–24, 2001, pp. 408–413.
35. Collin, W.D., The influence of slot geometry and segment spacing on diamond saw performance. Industrial Diamond Review, 1977, 37(1), 48–54.
36. Sakarcan, M., Wear resistant abrasive cutting wheel. U.S. Patent 4,854,295 (August 8, 1989)
37. Anon., Tensioning of a diamond blade. J. Chaland & Fils, Paris, France.
38. Pratt, W.R., Tension of diamond sawblades. In Proceedings of the 2nd DWMIT Technical Symposium Sawing and Grinding with Diamond Wheels, Chicago, USA, March 7, 1973, pp. S9/1–4.
39. Büttner, A., Mummenhoff, H., Testing the stress in diamond circular sawblades for sawing natural stones and concrete. Industrial Diamond Review, 1973, 33(5), 376–379.
40. Huelmann, E., Kohl, H.W., Optimum quality control of sawblade centres. Industrial Diamond Review, 1991, 51(6), 278–279.
41. Laforte, P.A., Whiting, R.R., Diamond blade sawing costs aren't set in granite. Finer Points, 1992, 4–6.
42. Haywood, R.A., Choosing and using diamond for stone sawing. Stone Industries, 1980, 16–19.
43. Bullen, G.J., Choosing the best grit for the job. Industrial Diamond Review, 1982, 42(1), 7–12.
44. Bailey, M.W., Bullen, G.J., Diamond sawing in the stone and construction industries. De Beers Industrial Diamond Division internal paper.
45. Bullen, G.J., Wright, D.N., Aspects governing diamond grit selection for stone and concrete applications. De Beers Industrial Diamond Division internal paper.
46. Ertingshausen, W., Wear processes in sawing hard stone. Industrial Diamond Review, 1985, 45(5), 254–258.
47. Konstanty, J., Theoretical analysis of stone sawing with diamonds. Journal of Materials Processing Technology, 2002, 123(5), 146–154.
48. Anon., Specifications for the use of diamond frame saw blades. Diamant Boart, Bruxelles, Belgium.
49. Anon., Case studies of diamond wire saw technology. GE Superabrasives, 1991. Booklet GES 90-960 E.
50. Hawkins, A.C., Brauninger, G., Diamond wire sawing for the construction and renovation industry. Paper presented at Superabrasives'91 Conference, Chicago, USA, June 11–13, 1991.
51. Davis, P.R., The future of diamond abrasives in stone processing. Industrial Diamond Review, 2001, 61(3), 159–167.
52. Wright, D.N., Engels, J.A., The environmental and cost benefits of using diamond wire for quarrying and processing of natural stone. Industrial Diamond Review, 2003, 63(4), 16–24.
53. Cai, O., Mancini, R., Diamond wire for cutting hard rock. Industrial Diamond Review, 1988, 48(5), 212–214.
54. Hawkins, A.C., Antenen, A.P., Johnson, G., The diamond wire saw in quarrying granite and marble. Dimensional Stone, 1990,
55. Tönshoff, H.K., Friemuth, T., Hillmann-Apmann, H., Diamond wire sawing of steel components. Industrial Diamond Review, 2001, 61(3), 203–208.

56. Anon., Diamond beads 'Turbo'. Acimm per il Marmo, 1993, 47(5), 149–151.
57. Bortolussi, A., Caranassios, A., Ciccu, R., Lassandro, R., Manca, P.P., Massacci, G., Progress in the knowledge of granite cutting with diamond wire. In Proceedings of the 11th International Conference on Ground Control in Mining. The University of Wollongong, N.S.W., July 1992, pp. 593–599.
58. Bortolussi, A., Ciccu, R., Manca, P., Massacci, G., Computer simulation of diamond-wire cutting of hard and abrasive rock. Transactions of the Institution of Mining and Metallurgy, Section A, 1994, 103, A124–A128.
59. Anon., Specifications for the use of diamond drilling tools. Diamant Boart, Bruxelles, Belgium.
60. Wright, D.N., Wapler, H., Investigations and prediction of diamond wear when sawing. Annals of the CIRP, 1986, 35(1), 239–244.
61. Asche, J., Diamond tools in the machining of natural stone – cutting principles, wear and future trend. Diamante Applicazioni & Tecnologia, 1999, 5(18), 61–67.
62. Tönshoff, H.K., Hillmann-Apmann, H., Asche, J., Diamond tools in stone and civil engineering industry: cutting principles, wear and applications. Diamond and Related Materials, 2002, 11, 736–741.
63. Custers, J.F.H., Plastic deformation of glass during scratching. Nature, 1949, 163(4171), 627.
64. Lutz, G., Grinding technical ceramics. Finer Points, 1995, 7(3), 6–30.
65. Malkin, S., From steels to ceramics in grinding. Finer Points, 1995, 7(3), 18–22.
66. Bianchi, E.C., da Silva, E.J., da Silva, C.E., de Souza, G.F., Fortulan, C.A., Fernandes, O.C., Valarelli, I.D., Aguiar, P.R., The behaviour of resin-bond diamond wheels in the grinding of advanced ceramics. Industrial Diamond Review, 1998, 58(4), 105–110.
67. Regiani, I., Fortulan, C.A., de Moraes Purquerio, B., Abrasive machining of advanced ceramics. Industrial Diamond Review, 2000, 60(1), 37–42.
68. Rübenach, O., From process innovation to product innovation – ultrasonic-assisted diamond turning of optical glass. Industrial Diamond Review, 2003, 63(4), 40–49.
69. Anon., Diamond sawblades and associated products. Triefus Industries Limited.

CHAPTER 3

Diamond Tool Design and Composition

The tool design and composition of its cutting layer play a key part in economic machining of materials where great demands are made on the tool life, its free-cutting ability and also on the precision of the job, edge quality and surface finish.

3.1 The Effect of Tool Design on its Performance

The tool design has been found of paramount importance especially in the case of circular sawing by means of segmental sawblades. The proper choice of slot geometry and arrangement of segments gives improved tool performance in terms of quality of the cut, noise, abrasive wear and fatigue life of the steel centre, flow of coolant to the cutting zone and reduced segment wear by abrasion.

Industrial experience has resulted in the adoption of several segmental sawblade shapes, which have been found to perform well in a wide range of applications. These basic shapes are shown schematically in Fig. 3.1.

Blades with narrow slots (see Fig. 3.1, type 1) are recommended for sawing hard materials, where good edge retention and surface finish have to be achieved.

By substituting each long segment with two, or more, shorter segments (type 2), it is possible to decrease the abrasive action of the slurry by shortening the trailing part of the segment and allowing greater flow of coolant into the cutting zone. Besides, the uneven spacing of segments, especially when the number of segments per tooth varies around the blade circumference, may slightly reduce the noise generated during

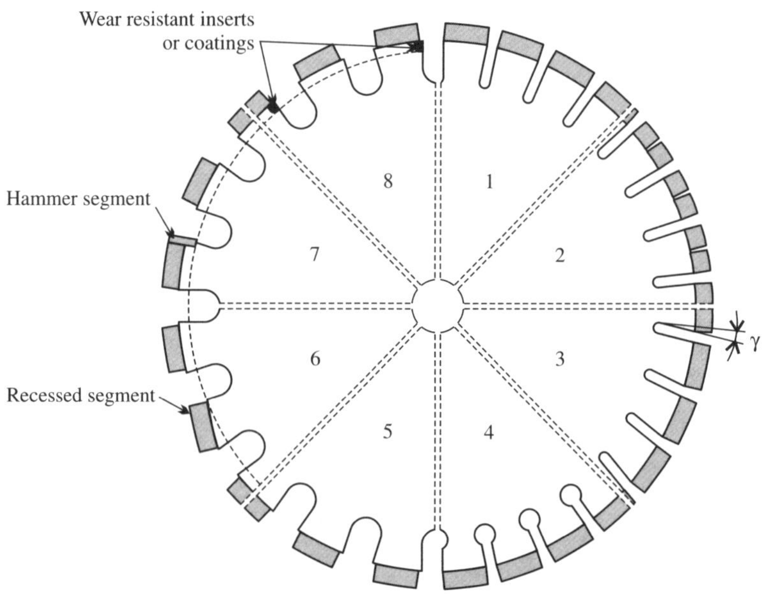

Figure 3.1 Typical periphery configurations of segmental sawblades.

cutting by eliminating natural resonant vibrations being induced by the segments contacting the workpiece at a regular frequency [1].

The application of a forward rake to each slot (type 3) or giving the slot a keyhole shape (type 4) have been found to hinder the fatigue crack initiation at the slot base thus having a beneficial effect on the life of the steel centre [1].

The fatigue life of the steel centre often becomes a critical consideration in sawblades used under heavy-duty conditions as well as in large diameter blades for stone sawing. In the first case, application of high cutting rates on hard materials give rise to increased cutting forces, which must be transmitted by the steel centre. In the second case, the largest steel centres are very expensive and have to be used a number of times to render the stone sawing process economically viable. Wide slots (type 5) are best suited for these applications since they effectively prolong the fatigue life of the centre.

Wider slots allow more efficient flushing of the swarf from the cut and, therefore, they are also readily used for fast sawing operations as well as in circular sawing of very abrasive materials. The presence of very abrasive particles between the sides of the blade and the workpiece may cause rapid side wear of the slotted rim leading to premature loss of segments. To prevent this, a number of wear protective segments (types 6 and 7) are incorporated into the blade periphery [2]. As each of these segments can effectively protect up to five trailing teeth, it has been found necessary to execute every sixth segment in this way. Alternatively the leading edge of, typically, every second slot has to be reinforced with abrasion resistant cemented carbide inserts or stellite coatings (type 8).

There are many ways in which diamond-impregnated segments may be designed. The basic classification of sawblade segments is given in Fig. 3.2, although various combinations of the types shown therein as well as many other less common executions [3–5] may also be found in the industrial practice.

Although uniform and simple-shaped segments are cheaper to manufacture, the application requirements may justify selection of complex designs.

The application of tapered segments has the advantage of decreasing the power consumed for sawing through reducing the lateral friction of segments against the stone.

There are two reasons for using segments incorporating a diamond-free base. First, they can hardly be worn down to the steel support and, therefore, it is cost-effective to withdraw the diamond from the base section of a segment. And second, the diamond-free layer of suitable fusion characteristics is required if the applied method of fixing the segment to the steel centre is by means of laser welding since a carbon-containing material would embrittle the steel support after welding.

The so-called 'sandwich' is a three-layer segment in which the outer layers differ from the inner one in their susceptibility to wear, whereas the multi-layer segments consists of a few diamond-impregnated layers separated by markedly thinner diamond-free zones [6,7]. Both types have been found superior to the conventional segments in the circular sawing applications due to desirable saddle-like or corrugated wear profiles being developed at work, as illustrated in Fig. 3.3, which impart a self-guiding characteristic to the sawblade and prevents it from deviating in the cut.

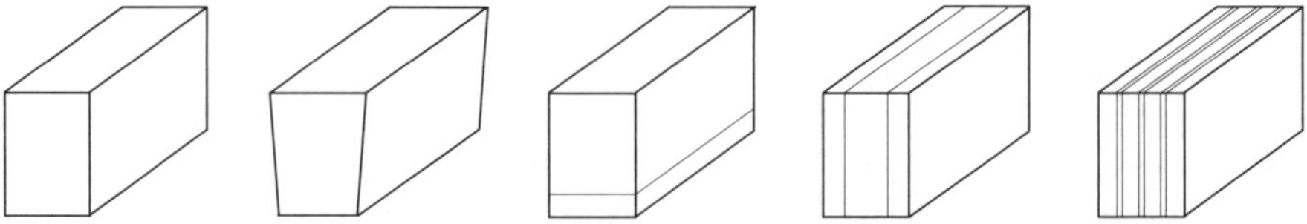

Figure 3.2 Basic types of sawblade segments.

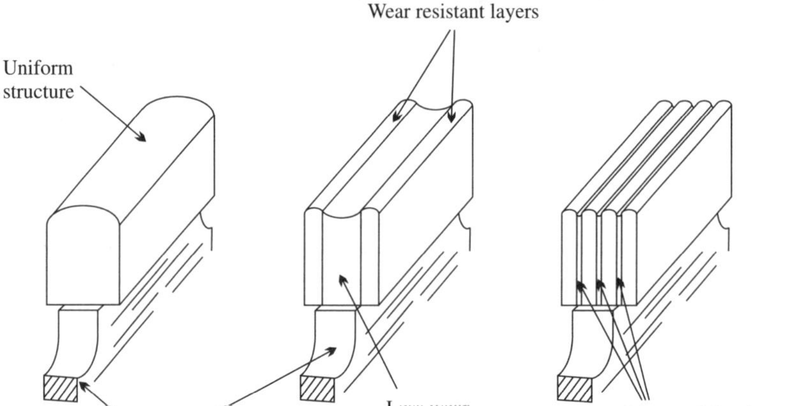

Figure 3.3 Wear patterns of conventional (left), sandwich (centre) and multi-layer type (right) diamond-impregnated segments.

3.2 Metal Matrix Selection

The two basic functions of the metallic matrix are to hold the diamond tight and to wear at a rate compatible with the diamond loss. The wear resistance of the matrix has to correspond with the abrasiveness of the workpiece material, so that neither the diamond grits protrude insufficiently nor they are lost prematurely. A too 'soft' a matrix wears faster than the diamond, which results in the possibility of diamond pullout. On the other hand, an extremely wear resistant matrix could wear more slowly than the break down of the diamond, causing the segment surface to polish. This phenomenon is commonly known as 'glazing'.

When a hard and dense material is machined, very little debris is formed and only as a fine powder. Under such conditions the matrix wear is slow. On the other hand, when an open-textured and gritty material is machined at high processing rates, a great quantity of coarse debris is released, which usually creates harsh wear environment for the matrix.

The workpiece machining conditions have also to be taken into consideration. During frame sawing, for example, the removal of detritus from the cut is hindered by the reciprocating motion of the sawblade. Hence, the wear rate of the matrix is always higher in comparison with circular sawing when the blade rotates in the same direction improving debris removal conditions.

The second and equally important function of the matrix is to hold the diamonds for as long as possible. The matrix capacity for retention of diamond grits is a complex, system-dependent property, which is affected by the material characteristics and application conditions as well.

The hold on the diamond may be mechanical or mechanical in combination with chemical. The potential retentive properties of the matrix are often assessed by its yield strength since to break the hold on a diamond particle the yield strength of the matrix has to be exceeded [8]. The pulling off force, however, is a function not only of the level of elastic deformation around the diamond but also of its shape, which can promote stress concentration. Hence the notch-sensitivity and ductility of the material are also important parameters [9,10].

Although in the vast majority of matrices the hold is purely mechanical, a great deal of effort has gone into improving retention by means of supportive chemical bonding. This can potentially be achieved by using alloys, which contain carbide-forming elements. Some of these alloys may melt at the hot pressing temperature to wet and flow over the diamond particles [11]. Consequently, islands of carbide nucleate on the diamond surface and grow with concentration of the reactive element to form a continuous interfacial layer. After this point is reached, the interface ceases to strengthen and may weaken as the carbide thickens due to the development of flows or porosity. Additionally severe diamond surface degradation is often observed due to the formation of too thick carbide layers. Since the optimum concentrations of carbide formers are far less than those needed to induce wetting, there are technical problems in utilising the most strongly bonding alloys as matrix materials.

To deal with these problems many attempts to bond diamonds in the solid state have also been made by examining various chromium [12,13] and tungsten [12,14] containing powders. On an industrial scale, however, this technique appears to be impracticable. The enhancement in the tool performance hardly ever recompenses for the additional powder treatment and necessity for graphite mould protection, as the mould also reacts with the powder being consolidated and may easily be destroyed during disassembling. Besides, the carbide forming elements impair ductility of the matrix. For instance, an excess of chromium in a cobalt-base matrix results in markedly increased porosity, due to the Kirkendall effect, or even formation of undesirable, brittle σ-phase [13]. For the above reasons, the recent trend is towards using metal-coated diamonds rather than chemically reactive matrices.

The segment manufacturing process determines the degree of diamond degradation. The properties of the diamond grit in the final product depend greatly on the segment processing temperature and chemical composition of the matrix. Synthetic diamonds begin to lose strength beyond 800°C. Due to the presence of metallic inclusions, the decline is fairly rapid at temperatures in excess of 1000°C. Therefore, the thermal stability of a diamond grit must be taken into consideration when the matrix powder requires hot pressing at temperatures of around 1100°C.

In addition to a drop in strength, surface graphitisation takes place for both synthetic and natural diamond grits. The process commences at ~700°C [15]. Obviously the percentage of diamond converted into graphite should be minimised, otherwise the tool life can be reduced. This is especially important in applications for which fine grit sizes are required. Generally, the finer the grit, the higher the temperature and the longer the hot stage processing time, the greater the degree of graphitisation. Metals showing appreciable solubility for carbon, e.g. iron, cobalt, nickel, etc., increase the possibility of severe surface attack and subsequent loss of diamond before it can be utilised [16,17].

Last but not the least, the higher the processing temperature the higher the consumption of hot pressing consumables, which often renders the production economically non-viable. As the hot pressing route is common in the manufacture of diamond-impregnated tool components, it is reasonable to classify the matrix powder mixtures according to the hot pressing temperature required to near-full densification of the segments. Such a classification based on the main segment fabrication limitations is provided in Table 3.1.

3.2.1 Cobalt Powders

Cobalt and cobalt-base alloys have been the most widely used as a base matrix material for diamond-impregnated tools for long. Unlike other metals, cobalt powder

- is available in a variety of grades, which differ in particle size and size distribution, particle shape, chemical purity, etc.,
- can be cost-effectively consolidated to virtually theoretical density, by means of hot pressing, under relatively low temperature/pressure conditions.

Table 3.1 Segment fabrication limitations.

Hot pressing temperature (°C)	Degree of diamond degradation		Consumption of hot press consumables (graphite moulds and electrodes, segment release agents, ceramic insulators, etc.)
	Grit type: low-grade synthetic (irregular, friable, with high content of metallic inclusions)	Grit type: high-grade synthetic (blocky, crystalline and thermally stable) or natural	
<900	Low	Low	Small
900–1000	Medium	Low	Medium[1]
1000–1100	High	Low to medium	High[1]

[1]Vacuum or inert gas atmosphere decreases consumption of graphite.

In a hot pressed condition, cobalt

- shows excellent diamond retention properties due to unique combination of high yield strength and toughness,
- has adequate resistance to abrasive wear, which can further be modified by admixing other powders prior to consolidation.

The experimental evidence show that the microstructure, phase composition, hardness, mechanical strength, ductility and resistance to wear of hot pressed cobalt vary widely with the powder properties and parameters of the hot pressing cycle [18–23]. The key factors that determine the choice of a particular powder grade include the mean particle size, and the type and content of impurities.

Particle size affects the densification temperature which, in turn, determines the degree of diamond degradation and, when in excess of 1000°C, causes excessive consumption of hot press consumables (see Table 3.1). With the air-operated equipment (see Fig. 4.6) this may render the process economically non-viable.

Under moderate pressure of 30–35 MPa, most of the commercial cobalt powders, listed in Table 3.2, require to be held for 2–3min at 700–900°C to achieve near-full density. As a rule, the coarser the powder, the more difficult it is to eliminate the residual porosity. The coarsest grades retain 4–5% of pores, even after hot pressing at 1000°C, which markedly degrade mechanical properties of the material.

In the tool-making practice cobalt powders coarser than 3 μm are readily mixed with bronze, which aids densification and renders the matrix suitable for processing less abrasive materials.

The majority of cobalt powders are made by chemical methods that have been reviewed elsewhere [23,25]. It is noteworthy that the powder's chemical, physical

Table 3.2 Commercial cobalt powders and their properties.

Grade designation	Fisher sub-sieve size (μm)	Typical impurities[1] (ppm)	Producer
SMS	~0.9	—	Umicore
Ultrafine	~0.9	Ag-940, S-45	
Extrafine	~1.4	—	
400 mesh	~4.5	Ca-290, Si-215, S-100	
CoUF	~0.9	Mg-250, Na-170, S-140	Eurotungstene
CoF	~1.25	n/a	
CoFS	~1.4	n/a	
CoC	~1.6	n/a	
CoD	~1.8	n/a	
CoH	~3.5	n/a	
Submicron	~0.8	—	OMG
Extrafine	~1.5	—	
400 mesh	~4	—	
Ultrafine	0.85–1.1	Mg-300, Na-130, Ca-100, Si-60	Sandvik [24]
Extrafine	1.15–1.5	Mg-200, Na-160, Ca-150, Si-60	

[1]Average values without oxygen.

and technological characteristics are specific to the type and purity of the precursor material and the production route.

Impurities such as oxygen, sulphur and stable oxide and sulphide formers, e.g. magnesium, calcium, sodium, etc., can strongly influence the development of microstructure with temperature thus affecting the post-consolidation properties of the material.

Oxides impede grain growth through anchoring the grain boundaries. In general, the higher the oxygen content of the powder the stronger the effect of oxide phase on the grain size stability with temperature. Consequently, the material resists recrystallisation far beyond its complete densification temperature and retains higher hardness and yield strength at the expense of ductility.

There is sufficient evidence that sulphur imparts brittleness to cobalt even when present in trace amounts of 50–60 ppm [26]. Sulphur may also stimulate grain growth at temperatures exceeding 877°C, due to formation of a liquid phase penetrating the grain boundaries and promoting their migration [23], which is typical of the Ultrafine hydrometallurgical powder from Umicore.

In the hydrometallurgical process, organic compounds are added to prevent growth and agglomeration of the cobalt particles [27]. This organic material forms a layer on the particle surface that minimises formation of metal oxides when the powder is stored in contact with air [28]. Therefore, the powder purifies itself during hot pressing to yield material virtually free from oxygen [23,28].

If powders are contaminated with magnesium, calcium, sodium and/or other elements, which form thermodynamically stable oxides and sulphides, then the material obtained thereof displays excellent grain size stability with the consolidation temperature and is not sensitive to sulphur-induced grain growth.

SEM micrographs of selected commercial cobalt grades used as matrix powders for diamond tools are shown in Fig. 3.4.

3.2.2 Cobalt Substitutes

The 1990's collapse of a strategic, Zairean cobalt supply led to serious concerns about cobalt's price stability [29]. Uncertainties facing supplies from Africa initiated an eager search for cheaper options. Consequently an extensive, market-oriented research on novel powders was triggered with the intention to substitute expensive and prone to unpredictable price fluctuations cobalt [30] with suitable alternatives.

The industry's evolving technological sophistication has led to the development of several low-cobalt and cobalt-free powders which consist of a combination of at least two elements, which are co-precipitated to yield pre-alloyed agglomerates of submicron-sized particles [31–33]. The manufacturing processes still undergo modifications to couple the product's excellent hot pressing characteristics with field performance similar to cobalt.

The chemical compositions and particle sizes of the cobalt substitutes have been compiled in Table 3.3.

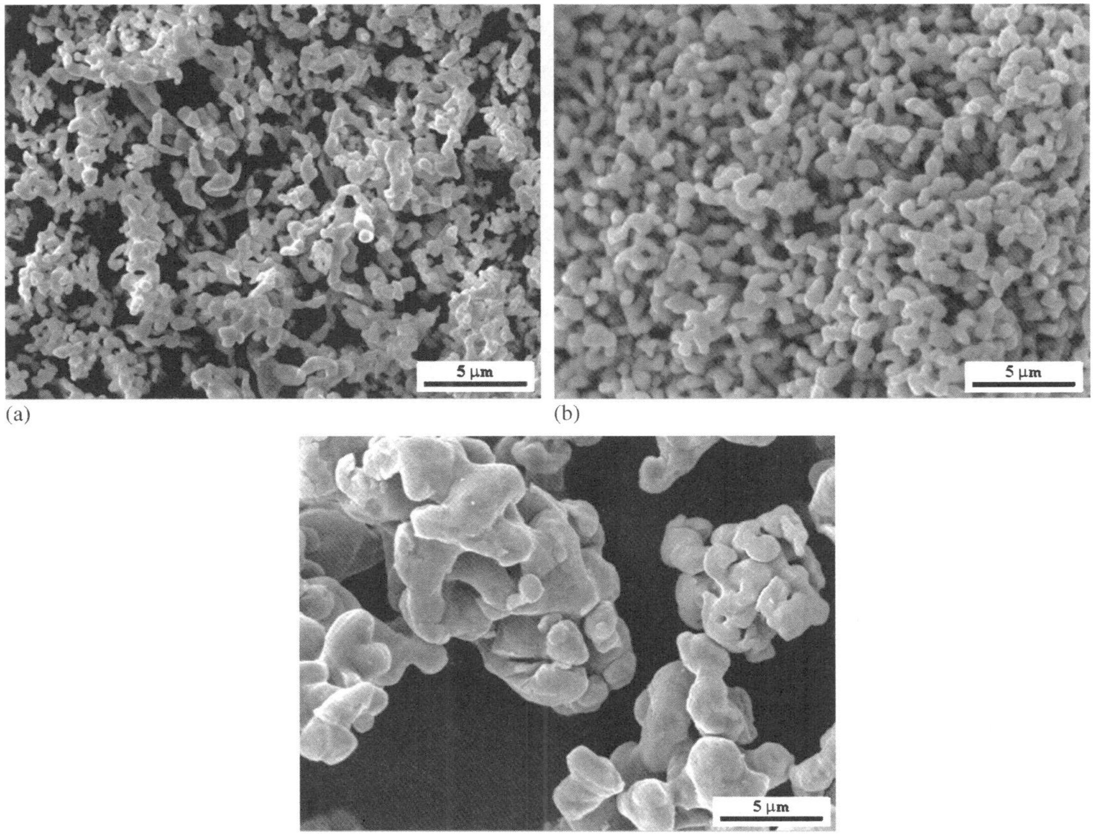

(a) (b) (c)

Figure 3.4 *Fine cobalt powders: (a) SMS (Umicore); (b) CoC (Eurotungstene); (c) 400 mesh (Umicore).*

Table 3.3 Typical chemical compositions and mean particle sizes of commercial cobalt substitutes.

Designation	Chemical composition[1] (wt.%)				Fisher sub-sieve size (μm)	Producer
	Fe	Cu	Co	Others		
Cobalite 601	70	20	10	—	~4.9	Umicore [33–36]
Cobalite HDR	66	7	27	—	6–7	
Cobalite CNF	68.4	26	-	Sn-3, W-2 Y_2O_3-0.6	~2	
Next 100	26	50	24	—	0.8–1.5	Eurotungstene [37–42]
Next 200	15	62	23	—	0.8–1.5	
Next 300	72	3	25	—	~4	
Next 900	80	20	-	—	~3	

[1]Oxygen not taken into account.

Typical morphologies of powders comprising the Cobalite and Next families are illustrated in Fig. 3.5.

3.2.3 Other Matrix Powders

Except in tools made by the infiltration process, the other powders are most commonly used as additives to cobalt, or its substitutes, with the intention to aid densification and to modify the matrix wear and retention characteristics. Iron, copper, tin, bronze alloys, tungsten, tungsten carbides and nickel alloys are used extensively.

Iron can be dangerously reactive with diamonds at high tool fabrication temperatures due to considerable solubility of carbon in austenite [14,15]. Nonetheless, moderate iron additions to cobalt (10 wt.% [23]) and Next range powders (20 wt.% [43,44]) markedly improve ductility of the matrix and, in the latter case, increase its yield strength [43,44]. There are many types of iron powders but mostly the finest grades have been utilised on a broader scale in diamond tool matrix compositions. Fine iron powder, prepared by thermal decomposition of iron pentacarbonyl [45], is illustrated in Fig. 3.6.

Additions of pre-alloyed bronze series powders, elemental copper and tin, used separately or in combination, usually assist densification of the matrix and reduce its resistance to wear thus rendering the tool more suitable for processing less abrasive materials [23].

Tin, pre-mixed tin bronzes and, in certain cases, pre-alloyed bronzes become molten at the hot pressing, or sintering, temperature. The amount of liquid, even if it appears as a transient phase, must be limited so that the parts retain their shape and minimum flash is forced out of the mould when the consolidation is performed by hot pressing. Besides the liquid phase strongly affects the degree of diffusion alloying between the matrix-forming components, therefore careful control over heating rate, time, temperature and pressure-at-temperature is required for reproducible results.

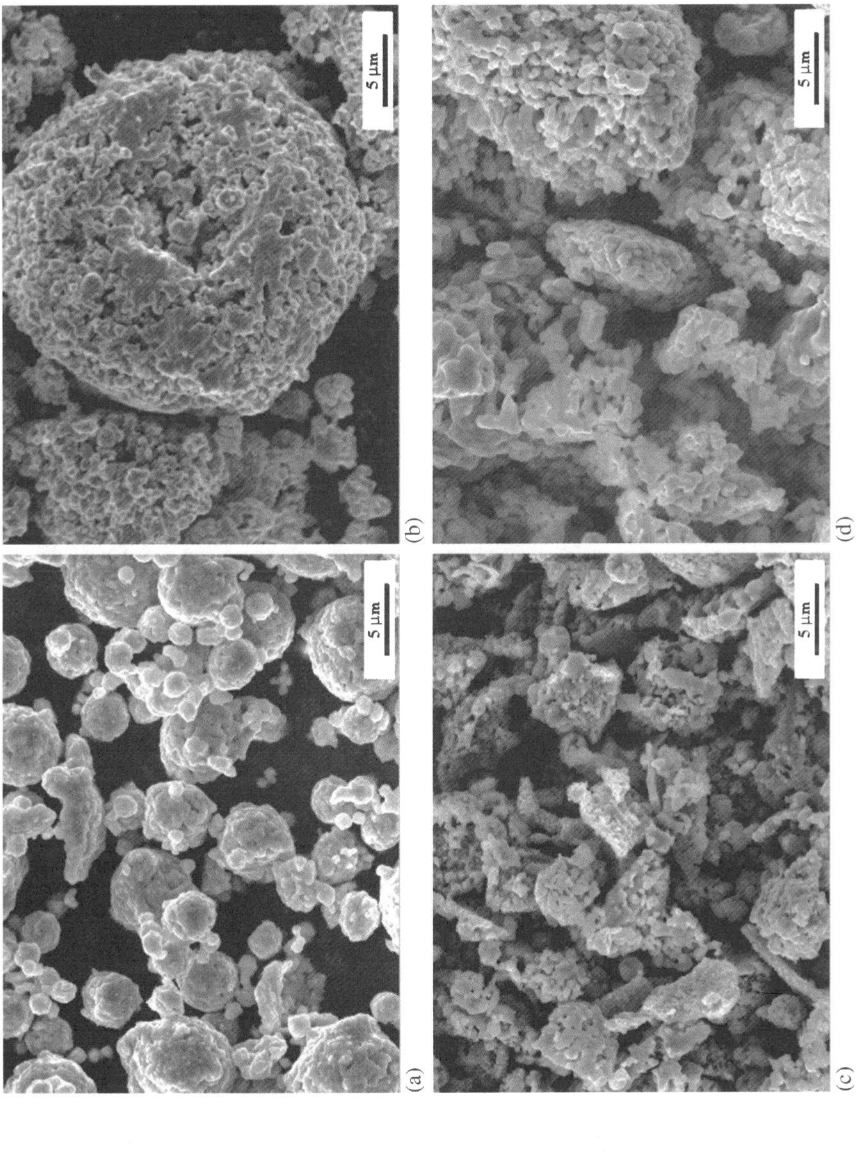

Figure 3.5 Selected powders used to replace cobalt: (a) Cobalite 601; (b) Cobalite CNF; (c) Next 200; (d) Next 900.

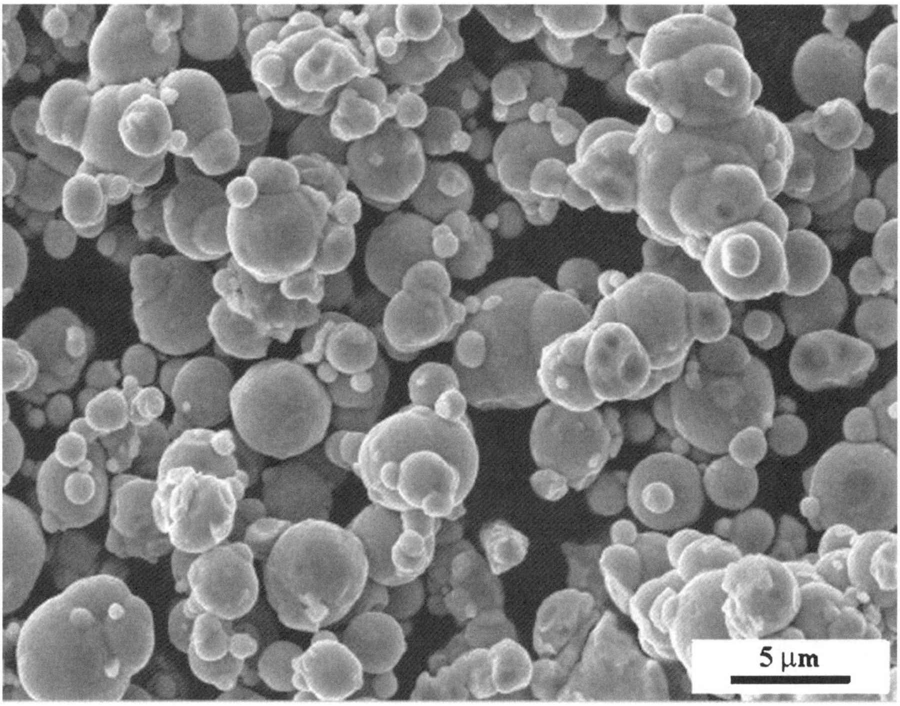

Figure 3.6 Carbonyl iron powder (Fe CN, BASF).

In the case of tin bronzes, larger amounts of either pre-mixed or pre-alloyed material are sometimes applied shifting the matrix composition to the bronze-base side. The various bronze formulation and consolidation conditions may largely affect the final material properties due to great complexity of microstructural modifications that can be produced. Figs. 3.7–3.9 compare morphologies of copper, tin and tin bronze powders typically used in the manufacture of diamond-impregnated tools.

Tungsten is one of the strongest and most rigid metals with strong atomic bonding affinity for carbon in the diamond, which results in the formation of a thin film of tungsten carbide at the interface between the tungsten and the diamond. Tungsten carbide does not decompose upon cooling and hence causes virtually no damage to the diamond [14]. These excellent characteristics may contribute to the firm anchoring of diamonds in the matrix, by providing both mechanical support and chemical bonding, as well as to the increased matrix resistance to wear.

The finest powders are commonly used in segments produced by the hot pressing route, whereas the coarser grades are better suited to the manufacture of diamond tools by infiltration. Fig. 3.10 shows a typical example of a fine tungsten powder.

Tungsten carbides are often used in various amounts for increasing the matrix resistance to abrasion. There are many types of tungsten carbide powders, which differ in the chemical composition, particle size, shape and microstructure.

Figure 3.7 *Electrolytic copper powder (FL, ECKA MicroMet).*

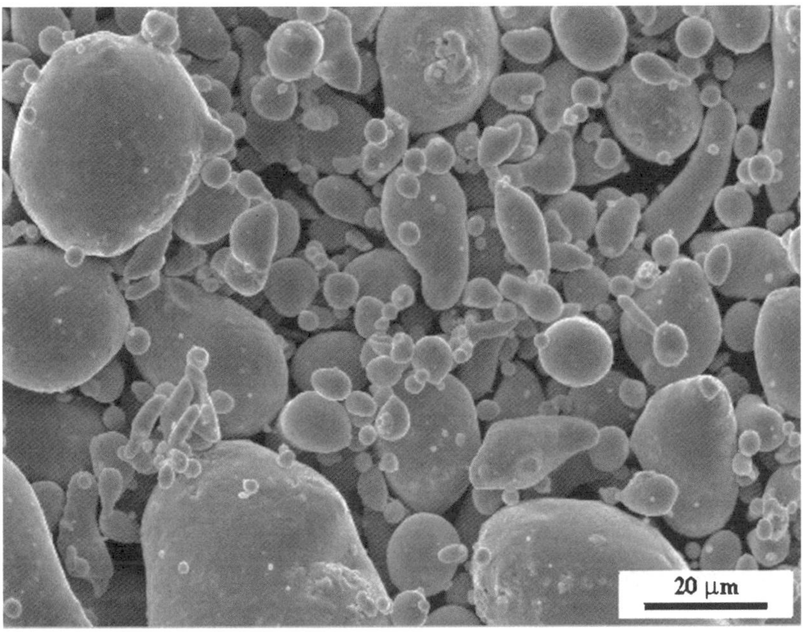

Figure 3.8 *Gas atomised tin powder (30 GN-350, ECKA Poudmet).*

Figure 3.9 Water atomised tin bronze powder (25 GR 85/15 -325, ECKA Poudmet).

Figure 3.10 Fine tungsten powder (HC 300, H.C. Starck).

The conventional-type fine tungsten monocarbide powder, produced by solid-state diffusion carburisation of tungsten metal powder at high temperature under protective atmosphere [46], is an irregularly shaped polycrystalline material characterised by a comparatively high particle porosity and variable particle microstructure [47]. Consequently, the powder exhibits poor flow and packing characteristics.

The so-called 'macrocrystalline' tungsten monocarbide is a low-porosity, non-agglomerated powder produced by a proprietary aluminothermic process, which consists in growing WC crystals from the tungsten ore concentrates via a self-sustaining exothermic reaction [48,49]. This process yields polygonal particles that may be up to two orders of magnitude larger than those produced from the traditional carburisation, as compared in Fig. 3.11.

Figure 3.11 Tungsten monocarbide powders most suitable for: (a) hot pressing/sintering (DS 250, H.C. Starck); (b) infiltration (Macrocrystalline WC, Kennametal).

The commercial macrocrystalline tungsten carbide powder is available in a wide range of sizes ranging from 1.2 to 420 μm [50]. Due to excellent wetting, flow and packing characteristics this powder is especially well suited to the manufacture of oil exploration and mining bits by the infiltration process. It may also be used to modify the wear properties of diamond-impregnated saw segments except in situations where its poor compactibility becomes critical.

Other types of tungsten carbide powders are two-phase eutectic WC/W_2C carbides. These powders are manufactured by either the traditional casting process [46,50] or by sintering in vacuum [46], and find applications similar to the macrocrystalline monocarbide. Although chemically similar to each other, the vacuum fused powder, shown in Fig. 3.12, is more isotropic and chemically uniform, and has lower porosity than its chill cast counterpart.

In coarse mesh sizes eutectic carbides are readily used to increase resistance to abrasion but may also act as a secondary abrasive in infiltrated oil and gas exploration bits. The finer sizes are better suited for the manufacture of tools by the hot pressing route that are utilised for sawing and drilling of abrasive materials.

An alternative method of improving the resistance to abrasion of the matrix is to add hard pre-alloyed nickel-base powders. A great number of gas atomised Ni–Cr–Si–Fe–B–C grades have been developed and made available on the commercial scale to date [51,52]. The powders are essentially finer than 150 μm and have near-spherical particle shape, as illustrated in Fig. 3.13.

Figure 3.12 Coarse eutectic tungsten carbide powder (KF110, Kennametal).

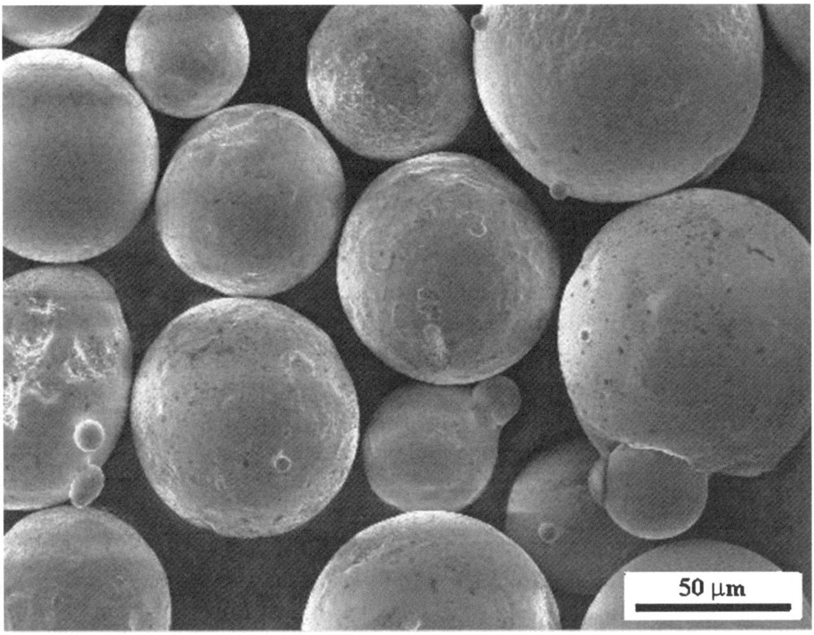

Figure 3.13 Nickel-base pre-alloyed powder (HMSP 1560-00, Höganäs).

The pre-alloyed nickel-base powders are less effective in aiding resistance to abrasion of the matrix than tungsten carbide powders. Important economic and technological advantages can, however, be achieved by substituting tungsten carbides with nickel-base alloys because they are cheaper, have lower specific density and, becoming soft at elevated temperatures, do not impair the hot compressibility of the matrix.

The other additive powders which are occasionally encountered in the tool-making practice comprise carbonyl nickel, atomised tool steels, manganese and manganese–nickel bronzes, tin-titanium bronzes, zinc, lead, molybdenum, boron, aluminium and chromium [53–57].

3.3 Diamond Grit Selection

To meet the individual requirements of a given application, a number of diamond-related parameters must be taken into consideration. The following have the greatest significance:

- diamond type,
- grit size,
- concentration.

3.3.1 Diamond Type

The processed material primarily determines the type of diamond abrasive used in the tool. A good general rule is the harder the workpiece, the stronger the diamond grit has to be selected.

Diamond manufacturers provide the toolmakers with a wide range of natural and synthetic grits of varying properties such as mechanical strength, thermal stability and matrix retention characteristics.

Natural grits, made by crushing mined diamond boart, are free from metallic inclusions and hence, they show excellent thermal stability. Natural diamond particles can retain their original mechanical strength even up to 1400°C, whereas synthetic grits begin to lose strength beyond 800°C [58]. From the matrix retention viewpoint, crushed crystals possess excellent bonding characteristics, which result from the many re-entrant surfaces. This makes the natural diamond advantageous for frame sawing of marble, limestone and other less hard stones. On the other hand, the irregular diamond surface impairs mechanical strength. In more rigorous applications, this drawback may successfully be dealt with by using rounded diamonds, which are designed and processed to cope with high-impact forces [59]. The rounded shape implies, however, that higher cutting forces are generated and hence rigid and more powerful machines are required. It is reasonable to choose rounded natural grits for manufacturing processes, which involve high hot pressing temperatures and give priority to low tool fabrication costs.

The two mentioned natural diamond abrasives are illustrated in Fig. 3.14.

The main advantage of using synthetic diamond is that it can be designed and manufactured to satisfy virtually any specific application requirements. In response to the diversification of the demands of the market, two major diamond grit families of different particle characteristics have been developed. By using either a cobalt- or a nickel-base alloy as a catalyst/solvent in the diamond synthesis crystals of different internal structure can be obtained. The 'cobalt' grades are characterised by the presence of ordered arrays of metal inclusions within each particle, whereas in the 'nickel' grades the impurities are uniformly distributed throughout the particle resulting in excellent transparency and cosmetic appearance and higher ability to retain mechanical strength after processing at high temperatures [60]. Fig. 3.15 exemplifies the difference between the two synthetic diamond families.

It has been well documented that the group VIII metals are responsible for catalytic graphitisation of diamond [61]. As transformation from diamond to graphite results in a volume increase of about 56%, the combination of graphitisation stresses and thermal stresses, originating from the mismatch in the thermal expansion coefficients at the diamond/catalyst metal interface, causes weakening or even shattering of the diamond particles [59]. Consequently, the amount and manner in which inclusions are distributed within each crystal in conjunction with its thermal history will affect the diamond's wear behaviour.

It has been observed in practice that grits manufactured by the cobalt alloy-based synthesis process, where inclusions are well aligned and ordered in a geometric pattern,

(a)

(b)

Figure 3.14 Natural diamond abrasives with (a) re-entrant surfaces and (b) rounded.

show a tendency to fracture in more irregular manner compared to their counterparts from the nickel alloy-based synthesis. This feature may either be beneficial to the tool, promoting a free-cutting behaviour at low material removal rates, or disqualify the tool for heavy-duty applications [60].

Over the past four decades, laboratory techniques have been developed for the quantification of the strength and thermal stability of diamond grits. The current industry standard procedure is an impact test, which consists in vibrating a cylindrical test capsule that contains the diamond sample and a test ball that impacts the diamond particles in the capsule [62–65]. The diamond may either be examined in the untreated state or after heat treatment at elevated temperatures [60,65]. The latter

(a)

(b)

Figure 3.15 Synthetic diamond abrasives produced by using (a) cobalt-base alloy and (b) nickel-base alloy as a catalyst/solvent in the diamond synthesis process.

procedure helps to predict the complex effects of segment thermal processing on diamond strength [58], although the matrix itself may accelerate the diamond degradation due to chemical reactions occurring at the diamond/matrix interface [16,17].

There are two ways of grit strength determination. The first is based on the half-life concept, i.e. the time required to reduce 50% of the diamond below the size of a breakdown sieve is determined. The second consists in measuring the percentage of diamond remaining on a breakdown sieve upon fixed comminution time, which is commonly termed the 'toughness index' (TI) or the 'thermal toughness index' (TTI) if the sample has additionally been annealed prior to testing.

Another technique, which has recently been developed, is the so-called 'compressive fracture strength' (CFS) test [66,67]. It uses two rotating rolls connected to a spring, load meter and data acquisition system to dynamically load and crush individual diamond crystals in a test sample. The force required to fracture each crystal is measured and processed to finally provide the average value as well as comprehensive information on the distribution of crystal strengths in a population.

Whichever method has been selected, it provides useful grit strength indicators that help the toolmaker to design the segment composition and fabrication process and also to predict the grit wear behaviour in application.

In addition to the grit strength indicators, the diamond manufacturers have developed means to define the diamond shape [67] as an equally important product quality. As exemplified in Fig. 3.16, the ideal crystal morphology of synthetic diamond can range from a pure cube to a pure octahedron.

The actual morphologies, created by the synthesis process conditions, lie between these two extremes and are termed 'cubo-octahedral'. As it is possible to assign the so-called 'morphology index' [68] or 'tau' parameter [67,69] to any diamond crystal, sophisticated image-analysis techniques have been developed to serve as a powerful tool in the quantification of particle morphology. Thereby the shape, which term encompasses some more structural features such as particle aspect ratio, its surface morphology, edge geometry, etc., may also be determined. Nowadays, highly specialised computerised systems are capable of capturing and processing extensive image analysis data to yield quantitative particle shape information as well as meaningful population statistics. Better shape recognition has a direct impact on the product consistency and also contributes to the creation of toolmaker-oriented, grit ranking systems [60].

The diamond shape itself will obviously affect the particle integrity and breakdown characteristics in use. Therefore diamonds which have a regular, cubo-octahedral shape are stronger than irregular crystals with less well-defined edges and rough faces. However, less crystalline grits are more free-cutting in application and show improved retention in the matrix. Since weaker diamond crystals with good retention properties are generally required in frame sawing, appropriately selected synthetics are seen as having potential to increasingly replace natural grits in this application.

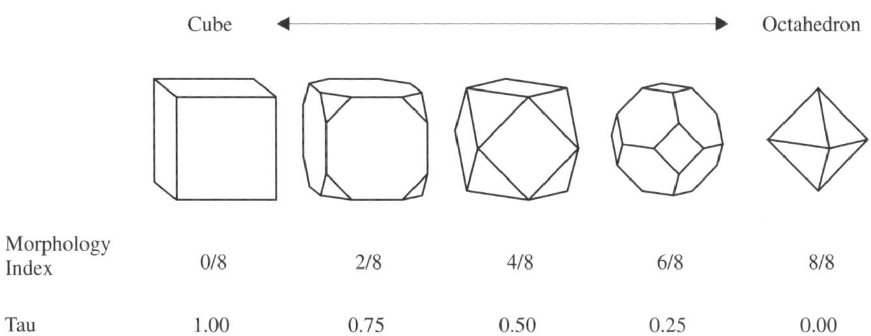

Figure 3.16 Diagrammatic representations and numerical codes of various synthetic diamond morphologies.

3.3.2 Grit Size

Commercial diamond abrasives, both natural and synthetic, are classified into two broad bands of grit sizes. These are commonly referred to as 'saw' and 'wheel' grits. Within each band the grit size is defined by two sieve aperture sizes, of the top and bottom defining sieves, as required by the existing standardised procedures [70]. Certain saw diamond grits are additionally characterised by a so-called 'particle per carat' (PPC) count [71]. The PPC specification guarantees the number of patricles per carat to a strict tolerance for each standard mesh size and thus aid in the product consistency.

The saw diamond abrasives are coarser than 80 US mesh and find extensive application in a variety of sawing, drilling and milling operations where fast material removal rates are essential. The general guidelines for diamond sizes used in the most typical tool applications are as follows [72–76]:

20/30 US mesh – circular sawing and drilling of very abrasive sandstone, fresh concrete and asphalt; drilling of reinforced concrete; frame sawing of abrasive sandstone.
30/40 US mesh – circular sawing and drilling of stone, concrete, reinforced concrete and asphalt; frame sawing of abrasive sandstone.
40/50 US mesh – circular sawing, drilling and milling of stone (e.g. granite, diorite, gabbro, less abrasive sandstone, limestone, dolomite, marble), concrete and refractories; wire sawing of stone and construction materials; calibrating of stone slabs.
50/60 US mesh – circular sawing and drilling of very hard fine-grained igneous stone (e.g. granite, granodiorite) and refractories; circular sawing, frame sawing and drilling of fine-grained sedimentary and metamorphic stone (e.g. limestone, marble); calibrating of stone slabs.
60/80 US mesh – circular sawing, drilling and milling of glass; calibrating of ceramic tiles.

There is a general tendency to apply finer grits for slow, secondary sawing operations where perfect surface finish and edge definition are matters of great concern. In any brand of diamonds the finer grits, being relatively stronger, are also recommended for sawing harder and more difficult to cut materials.

The main advantage of using coarser grit is a bigger potential for faster cutting rate. This is because a coarse grit attains higher protrusions, and hence produces greater clearance, and enables easier swarf ejection when a larger volume of workpiece material has to be removed per tool revolution. However, if too coarse grit is used on very hard materials, the penetration of each particle is limited and either excessive pullout occurs or large wear flats appear on the diamonds [77].

Diamond abrasives finer than 80 US mesh fall in the wheel range and are generally used for grinding operations. Most commercial products are available as standard in a full complement of nine sieve sizes, from 80/100 down to 325/400 US mesh [78–80], to meet the requirements of the consecutive grinding steps, which eliminate the marks left by the previous surfacing operations. Thus, a calibrated, or sawn, surface can be gradually changed to a semi-finished surface capable of being polished.

3.3.3 Concentration

The amount of diamond in a segment or bead is based on a scale in which 100 concentration is equivalent to 4.4 carats/cm^3 (25 vol%). All other concentrations are proportional.

The diamond concentration, in conjunction with the diamond particle size, governs the number of cutting points per unit area of the working face of the tool. As demonstrated in Table 3.4, the total number of diamonds and pullout sites is proportional to the concentration and inversely proportional to the grit size.

Which concentration to use depends on a number of factors. When designing the tool composition, the material to be processed and its properties should be considered first. In general, the easier to cut and more abrasive the workpiece, the higher should be the diamond concentration.

When applying high concentration to increase the tool life, a sufficiently high machine power must be available. Otherwise the load on an individual grain is correspondingly low, i.e. the grain penetration depth is shallow, which results in polishing the diamond. In such a case, little or no cutting action is observed and the tool requires redressing.

On lowering the diamond concentration, the load on an individual diamond becomes comparatively high, increasing the amount of material removed per grit particle until a certain critical point is reached. When the grit strength/toughness is exceeded the diamonds start to shatter or are pulled out of the matrix prematurely.

It is interesting to note that only a relatively small proportion, i.e. about 24–26% [82,83], of protruding diamonds has the correct height of protrusion to actively work at any given time.

Table 3.4 Total number of diamond crystals and pullout sites per square centimetre of surface area on a worn segment [81].

US mesh size	Diamond concentration					
	15	20	25	30	35	40
25/35	14	19	24	29	34	38
30/40	22	30	37	45	52	60
35/40	26	34	43	51	60	68
40/50	38	51	63	76	88	101
50/60	65	87	109	131	153	174
60/80	85	114	142	170	199	227

3.3.4 Coated Grits

As already mentioned (see Section 3.2), the very important role of a matrix in diamond-impregnated segments is to hold the diamond crystals for as long as possible. In general, the proportion of each diamond particle which performs work is small when compared with that portion which is lost because the matrix is unable to retain it. Figures lower than 25% are considered normal [84], because the vast majority of metallic matrices rely on purely mechanical retention of the diamond. In particular, it is not easy to develop a matrix capable of effective utilisation of very strong synthetic grits, characterised by perfectly defined blocky shape and smooth faces. Therefore, the diamond itself has to co-operate with the matrix so as to minimise its excessive pullout.

As illustrated schematically in Fig. 3.17a, during deformation of the matrix, which gives support to a loaded diamond, a slip of metal along the face of the diamond occurs. The static coefficient of friction for diamond on metals is very small and ranges between 0.1 and 0.15 [85]. Therefore the metal is nearly free to slip laterally at the interface. The situation changes dramatically when the coefficient of friction is high. Growing frictional forces increase the pressure necessary to produce plastic yielding thus strengthening the matrix under the diamond [23,86,87]. In the extreme case, as exemplified in Fig. 3.17b, strong adhesion of the matrix to the diamond surface compels slip to take place within the skin of the metal near the diamond rather than at the interface itself.

In the tool-making practice, there are two possible ways whereby the diamond-to-matrix bonding strength can markedly be improved by increasing friction at the diamond–matrix interface. The first method consists in thermal or chemical treating the diamond grains with the aim to roughen their surface [88–90]. The known processes have, however, an essential disadvantage that a relatively large quantity of material must be removed from the diamond surface before its topography guarantees

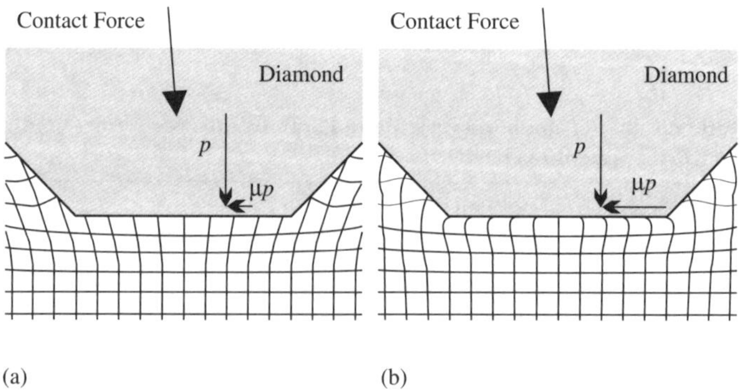

Figure 3.17 Effect of friction on deformation of the matrix by a diamond crystal: (a) local combination of normal pressure p and coefficient of friction μ produces frictional force that counteracts the lateral slip of material; (b) strong chemical bonding between matrix and diamond causes slip to occur at certain distance from the interface.

satisfactory improvement in the bonding strength. There is also a considerable risk that such surface etching may harm the diamond's integrity and strength. Therefore, in the case of the saw diamond grits, this method has not found broader application.

Much more effective way to improve diamond retention in the matrix leads through coating the crystals with a thin film of strong carbide former such as titanium (Fig. 3.18a) [84,91–99], chromium [92–97] or silicon [100]; but zirconium [92], tungsten, tantalum, molybdenum, niobium or alloys thereof [101] may also be applied. An additional layer [84,93,101,102], or layers [92], preferably nickel-based, may optionally be applied with the intention to protect the carbide former from oxidation, to render the material 'compatible' with the matrix material, and to minimise diamond pull out in soft matrices by additional mechanical interlocking provided by a spiked morphology of the coating (Fig. 3.18b) [102].

(a)

(b)

Figure 3.18 Diamond abrasives coated with (a) a single layer of titanium carbide (SDBTCH, Element Six), and (b) an additional layer of nickel alloy (MBS SB, GE).

The most suitable methods for deposition of the primary carbide-forming layer produce strong chemical bonding with limited exposure to harmful thermal cycles. Various chemical vapour deposition (CVD) techniques are most preferred [93,97,101] but other methods, such as metal vapour deposition [92,103] or packed salt cementation [93], may also be used.

The primary carbide layer is relatively thin (0.1–10 μm) but the secondary layer is usually applied in an amount of between 20 and 60 wt.% of the abrasive grits [93,102]. Therefore, the total weight of the coating must be precisely known and taken into account during the preparation of the matrix–diamond mixture.

The benefits to be obtained when using coated diamonds are: improved tool life [93–99], especially significant at low blade peripheral speeds [84] and at high cutting rates [98,99,104]; diamond protection from surface graphitisation, oxidation and attack by aggressive matrix components [91,93–99]; and strengthening of flowed diamonds through 'healing' surface defects [104].

The extended tool life is apparently attributable to the increased number of working crystals and their increased protrusion, as well as to enhanced transmission of dynamic strain to the tool shank which may further inhibit premature cutting point failure [86,87]. It should be emphasised, however, that improper tool manufacturing conditions often give rise to coating failure and has a detrimental effect on tool performance in application. The most serious hazards are: oxidation of the coating due to application of highly oxidised matrix powders; dissolution of the coating in the matrix which may occur during prolonged exposure to too high temperature during the hot pressing or sintering process [98].

Although, earlier the use of coated grits was mainly restricted to process granite [94], the current trend is towards developing titanium-base alloy coated medium-to-low grade grits for sawing less demanding materials such as marble, limestone, slate, sandstone, etc. [96,97,100,105].

References

1. Collin, W.D., The influence of slot geometry and segment spacing on diamond saw performance. Industrial Diamond Review, 1977, 37(1), 48–54.
2. Wöll, E., Cutting asphalt with diamonds. Diamonds in Industry – Roads and Runways, De Beers Industrial Diamond Division, London, 1984, pp. 44–51.
3. Hausberger, P., Kerber, B., Cutting tool. International Patent Application PCT/AT 91/00088 (July 23,1991).
4. Dietel, K., Segmented cut-off blades with polygon technology. Industrial Diamond Review, 2002, 62(3), 173–176.
5. Reinhardt, K.-D., New shape of diamond segment for machining natural and artificial stone. Industrial Diamond Review, 2003, 63(2), 16–18.
6. Yang, K.H., Pan, B.S., Duan, L.C., Theoretical determination and experimental verification of the structure of layered segment of diamond saw blade. In Machining of Natural Stone Materials, edited by Xu, X., Trans Tech Publications Ltd, Uetikon-Zuerich, Switzerland, 2003, pp. 99–102.
7. Zhan, Z., Tang, J., A kind of particular segment with tapered shape and multi-layers for diamond saw blades. Paper presented at Intertech 2003 Conference, Vancouver, Canada, July 28–August 1, 2003.

8. de Châlus, P.A., Metal powders for optimum grain retention. Industrial Diamond Review, 1994, 54(4), 170–172.
9. Chalkley, J.R., Thomas, D.M., The tribological aspects of metal-bonded diamond grinding wheels. Powder Metallurgy, 1969, 12(24), 582–597.
10. Konstanty, J., Developing a better understanding of the bonding and wear mechanisms involved in using diamond impregnated tools. In Proceedings of International Workshop on Diamond Tool Production, Turin, Italy, November 8–10, 1999, pp. 97–106.
11. Scott, P.M., Nicholas, M., Dewar, B., The wetting and bonding of diamonds by copper-base binary alloys. Journal of Material Science, 1975, 10, 1833–1840.
12. Levin, E., Gutmanas, E.Y., Solid-state bonding of diamond to Nichrome and Co-20wt.% W alloys. Journal of Materials Science Letters, 1990, 9(6), 726–730.
13. Akyüz, D., Hofmann, H., Interface aspects in cobalt-based diamond cutting tool segments. In Proceedings of Powder Metallurgy World Congress & Exhibition, Granada, Spain, October 18–22, 1998, Vol. 4, pp. 158–163.
14. Anon., The metallurgy of diamond tools. Industrial Diamond Review, 1985, 45(5), 248–250.
15. Young, B., The graphitisation of diamond during the manufacture of diamond tools. Diamond Information L2, De Beers Industrial Diamond Division, London.
16. Tuzzeo, J.J., Bovenkerk, H.P., Ratterman, E., Effective utilization of synthetic diamond in high temperature metal bonds. General Electric Company publication, Worthington, Ohio.
17. Bullen, G.J., The effect of temperature and matrix on the strength of synthetic diamond. Industrial Diamond Review, 1975, 35(5), 363–365.
18. Konstanty, J., Bunsch, A., Hot pressing of cobalt powders. Powder Metallurgy, 1991, 34(3), 195–198.
19. Konstanty, J., Bunsch, A., Cias, A., Factors affecting hardness and ductility of hot-pressed cobalt powders. Powder Metallurgy International, 1991, 23(6), 354–356.
20. de Chalus, P.A., Konstanty, J., Diamond tooling – stone cutting. Cobalt News, 1996, (4), 12–17.
21. Konstanty, J., Hot compacting of extrafine cobalt. In Advances in Powder Metallurgy & Particulate Materials – 1996, edited by Terry M. Cadle & Narasimhan, K.S., Metal Powder Industries Federation, Princeton, NJ, 1996, Vol. 2, Part 5, pp. 3–7.
22. Konstanty, J., Cobalt and diamond tooling. In Proceedings of the Cobalt Conference, Hong Kong, April 23–24, 1997, Presentation No. 5, pp. 1–10.
23. Konstanty, J., Cobalt as a Matrix in Diamond Impregnated Tools for Stone Sawing Applications. AGH Uczelniane Wydawnictwa Naukowo-Dydaktyczne, 2nd Edition, Krakow, 2003.
24. Cobalt. Sandvik Asia Ltd., Product information brochure, January 2002.
25. Konstanty, J., Production of cobalt powders for cemented carbide and diamond tool industries. Archives of Metallurgy, 2002, 47(1), 43–56.
26. Young, R.S. (Ed.). Cobalt. Its Chemistry, Metallurgy, and Uses. Reinhold Publishing Corporation, New York, Chapman & Hall, London, 1960.
27. Anon., Hydrometallurgically processing fine cobalt. Metal Powder Report, 1996, (12), 18–22.
28. Hardmetal Technology. Oxygen effects. Union Miniere. Fact sheet, Issue 1, August 1998.
29. Clark, B., Cobalt and hard metal. In Advances in Powder Metallurgy & Particulate Materials – 1996, edited by Terry M. Cadle & Narasimhan, K.S., Metal Powder Industries Federation, Princeton, NJ, 1996, Vol. 5, Part 18, pp. 3–7.
30. Hawkins, M., Trends in the supply and demand for cobalt. Cobalt News, 2003, (1), 10–14.
31. Standaert, R., Pre-alloyed, copper containing powder, and its use in the manufacture of diamond tools. U.S. Patent 6,312,497 (November 6, 2001).
32. Clark, I.E., Kamphuis, B.J., Recent advances in prealloyed powders for diamond tooling. In Diamond Tooling Proceedings of the European Conference on Hard Materials and Diamond Tooling, Euro PM 2002, Lausanne, Switzerland, October 7–9, 2002, pp. 35–42.
33. Clark, I.E., Kamphuis, B.J., Cobalite HDR – a new prealloyed matrix powder for diamond construction tools. Industrial Diamond Review, 2002, 62(3), 177–182.
34. Metal matrix powders for diamond tools. Union Miniere, Diamond Tool Brochure – DT/All Products/E/0998.

35. Cobalite. New times, new solutions. Union Miniere, Product information brochure, July 1997.
36. Kamphuis, B.J., Serneels, A., Cobalt and nickel free bond powder for diamond tools: Cobalite CNF. Industrial Diamond Review, 2004, 64(1), 26–32.
37. del Villar, M., Muro, P., Sánchez, J.M., Iturriza, I., Castro, F., Consolidation of diamond tools using Cu–Co–Fe based alloys as metallic binders. Powder Metallurgy, 2001, 44(1), 82–90.
38. Next 100. Eurotungstene Metal Powders. Technical data sheet, 05/2003.
39. Next 200. Eurotungstene Metal Powders. Technical data sheet, 04/2004.
40. Next 300. Eurotungstene Metal Powders. Product information brochure, July 2000.
41. Next 300. Eurotungstene Metal Powders. Technical data sheet, 03/2004.
42. Next 900. Eurotungstene Metal Powders. Technical data sheet, 03/2004.
43. Bonneau, M., A revolutionary binder for the diamond tool industry. In Proceedings of International Workshop on Diamond Tool Production, Turin, Italy, November 8–10, 1999, pp. 47–55.
44. Bonneau, M., Next and Next premixed powders: a complete range of basis. Diamante Applicazioni & Tecnologia, 1999, 5(18), 45–52.
45. Ebenhoech, F.L., Schlegel, R., Preparation of iron powder. U.S. Patent 4,652,305 (March 24, 1987).
46. Matrix powders for diamond tools. Kennametal Inc. catalogue, 1989.
47. Tungsten carbide (WC). Macro Division of Kennametal Inc. Technical data sheet IC-9-80-6.
48. Terry, C.J., Frank, J.D., Macrocrystalline tungsten monocarbide powder and process for producing. U.S. Patent 4,834,963 (May 30, 1989).
49. Terry, C., Morris, J., Macrocrystalline thermit process revealed. Metal Powder Report, 1999, (12), 22–27.
50. Findeisen, E., Frank, K., Becker, W., Muller, F., Process of manufacturing cast tungsten carbide spheres. U.S. Patent 5,089,182 (February 18, 1992).
51. Höganäs gas atomised powder. Höganäs AB, Product Information Brochure – HF 96-1.
52. Nickel base diamond bonding powders. SCM Metal Products, Technical Data - Bulletin No. 95638 (August 1, 1984).
53. Diamond bonding & abrasives – matrix powders. SCM Metal Products, Technical Data – Bulletin No. 95636, (June 21, 1984).
54. Sintering metal powder for the diamond tool industry. Dr. Fritsch KG, Product Catalogue DrF-D010196.
55. Foxmet S.A., Product Catalogue.
56. Pometon powders for diamond tools. Pometon S.p.A., Product information brochure 06/2002.
57. Diamond tool binders. Ecka Granules, Product information brochure 04/03.
58. Zsolnay, L.M., Selection of diamonds for segmental saws. Industrial Diamond Review, 1977, 37(6), 382–384.
59. Dyer, H.B., Conradi, V.R., Properties of natural Debdust and its use in diamond sawblades. Industrial Diamond Review, 1972, 32(4), 335–343.
60. Davis, P.R., Fish, M.L., Peacock, S., Wright, D.N., An indicator system for saw grit. Industrial Diamond Review, 1996, 56(3), 78–87.
61. Clifton, P.H., Evans, S., XPS studies of diamonds/transition-metal interfaces. Diamante Applicazioni & Tecnologia, 1997, 3(10), 38–49.
62. Belling, N.G., Dyer, H.B., Impact strength determination of diamond abrasive grit. The Industrial Diamond Information Bureau, London, 1964.
63. Belling, N.G., Bialy, L., The Friatester-10 years later. Diamond Information L33, De Beers Industrial Diamond Division.
64. Belling, N.G., Friatesting and diamond strength: a review. Industrial Diamond Review, 1992, 52(3), 133–137.
65. FEPA standard for measuring the relative strengths of saw diamonds. FEPA, Edition 1, May 30, 1994.
66. Anon., The future of diamond characterization. GE Superabrasives, 1993. Booklet GES 93–1019.
67. Anon., Diamond characterization. GE Superabrasives, 2000. Booklet GES 1278 E.
68. Bailey, M.W., Hedges, L.K., Crystal morphology identification of diamond and ABN. Industrial Diamond Review, 1995, 55(1), 11–14.

69. Bräuninger, G., Production and properties of synthetic diamond grit. In Proceedings of International Workshop on Diamond Tool Production, Turin, Italy, November 8–10, 1999, pp. 3–13.
70. A review of diamond and CBN sizing & standards. Industrial Diamond Association of America, Inc., Skyland, USA, Publication No. S&S593 5M.
71. Engels, A., The role of Particle Per Carat in diamond tool behaviour. Industrial Diamond Review, 2003, 63(2), 39–45.
72. Wilson, S.M., The effect of diamond size on drill bit performance when drilling reinforced concrete. De Beers Industrial Diamond Division internal paper.
73. Anon., De Beers diamonds in the glass industry. De Beers Industrial Diamond Division, Special Publications Series.
74. Cooper, D.G., The theory and application of diamond blades to refractory cut-off. In Proceedings of 2nd DWMIT Technical Symposium Sawing and Grinding with Diamond Wheels, Chicago Illinois, USA, March 7, 1973, pp. S7/1–4.
75. Burgess, R.R. Man-made diamond for stone processing. Paper presented at 1st Technical Symposium, Bucharest, Romania, October 5–6, 1978.
76. Garrard, R., Peacock, S.R., Hori, M., Pearce, N.R., The future role of diamond in the construction industry. Industrial Diamond Review, 2001, 61(2), 121–129.
77. Jennings, M., Wright, D., Guidelines for sawing stone. Industrial Diamond Review, 1989, 49(2), 70–75.
78. Premadia. Diamond abrasives wheel sizes. De Beers Industrial Diamond Division. Technical data leaflet 3.3.4.
79. RVG diamond products. Real value grinding. GE Superabrasives, 2002. Product information brochure GES 1320 E.
80. IMD Iljin metal bond diamond series. Iljin Diamond. Product information brochure.
81. MBS diamond products for sawing and drilling applications. GE Superabrasives, 1991. Product information brochure GES 91-966.
82. Wright, D.N., Wapler, H., Investigations and prediction of diamond wear when sawing. Annals of the CIRP, 1986, 35(1), 239–244.
83. Wright, D.N., Wilson, S.M., Brown, W.F., Ovens, U., Segment wear on diamond impregnated mining bits. Industrial Diamond Review, 1990, 50(5), 248–252.
84. Bailey, M.W., Collin, W.D., Investigations into diamond sawing using Titanized grits. Stone Industries, 1977, 18–21.
85. Bowden, F.P., Tabor, D., The Friction and Lubrication of Solids. The Claredon Press, Oxford, 1954.
86. Webb, S.W., Diamond retention in sintered cobalt bonds for stone cutting and drilling. Diamond and Related Materials, 1999, 8, 2043–2052.
87. Webb, S.W., Crystal retention improves ROI and performance of diamond tools. Paper presented at Intertech 2000 Conference, Vancouver, Canada, July 17–21, 2000.
88. Tokura, H., Yoshikawa, M., Heat treatment of diamond grains for bonding strength improvement. Journal of Materials Science, 1989, 24, 2231–2238.
89. Borse, D., Process for treating diamond grains. US Patent 5,035,771 (July 30, 1991).
90. Musu-Coman, M., Fecioru, M., Baluta, G., Arnici, N., Surface processing of ultrahard materials used for embedding in resin or metallic matrices. In Proceedings of Powder Metallurgy World Congress & Exhibition, Granada, Spain, October 18–22, 1998, Vol. 4, pp. 234–239.
91. Bullen, G.J., Wright, D.N., Aspects governing diamond grit selection for stone and concrete applications. De Beers Industrial Diamond Division internal paper.
92. Chen, S.-H., Sung, C.-M., Multiple metal coated superabrasives grit and methods for their manufacture. US Patent 5,024,680 (June 18, 1991).
93. McEachron, R., Connors, E.J., Slutz, D.E., Multi-layer metal coated diamond abrasives with an electrolessly deposited metal layer. US Patent 5,232,469 (August 3, 1993).
94. Brauninger, G., Hayden, S.C., MBS-960Cr2 and MBS-960Ti2 chromium- and titanium-coated manufactured diamonds for sawing and drilling applications. GE Superabrasives, 1995, Product information brochure GES 95-1093.

95. MBS-960 diamond titanium and chromium coated products. GE Superabrasives, 1995, MBS coated diamond product sheet GES 95-1072.
96. The MBS 900 diamond series coated crystals. GE Superabrasives, 2000, Product information brochure GES 1281 E.
97. Dey, N., Meng, Y., Titanium chromium alloy coated diamond crystals for use in saw blade segments and method for their production. US Patent 6,319,608 (November 20, 2001).
98. Engels, J.A., Egan, D., Using coated diamonds in diamond impregnated tools. In Proceedings of Euro PM2003 Conference, Valencia, Spain, October 20–22, 2003, Vol. 1, pp. 175–181.
99. Anon., SDBTC – coated saw grits from Element Six. Industrial Diamond Review, 2004, 64(1), 66.
100. The MBS 900 diamond series coated crystals. GE Superabrasives, 2002, Product information brochure GES 1326 E.
101. Chen, S.-H., Sung, C.-M., Chemically bonded superabrasive grit. US Patent 5,062,865 (November 5, 1991).
102. MBS SB diamond. GE Superabrasives, 2003, Product information brochure GES 1332 E.
103. Wang, Y.H., Zang, J.B., Wang, M.Z., Zheng, Y.Z., Relationship of interface microstructure and adhesion strength between Ti coating and diamond. In Machining of Natural Stone Materials, edited by Xu, X., Trans Tech Publications Ltd, Uetikon-Zuerich, Switzerland, 2003, pp. 41–45.
104. Hayden, S., Metal-coated diamonds boost tool performance in stone sawing and drilling. Dimensional Stone, September 1997.
105. Anon., The success continues ... MBS CMD diamond superior coating technology for marble, limestone and similar materials. Diamante Applicazioni & Tecnologia, 1999, 5(17), 128–131.

CHAPTER 4

Diamond Tool Fabrication

A typical fabrication process consists of two separate stages. The diamond-impregnated tool components are initially produced by the PM route and then, in the final stage, they are attached to the tool support which may require further processing.

In certain cases, however, e.g. in the manufacture of continuous rim sawblades or infiltrated drill bits, these two stages are combined in a single process in which the consolidation of the diamond-impregnated layer takes place simultaneously with cementing it to the tool shank.

4.1 Powder Metallurgy

The production route for all metal-bonded tool components involves a number of PM operations before the consolidated products are inspected for quality. The major manufacturing steps are shown diagrammatically in Fig. 4.1.

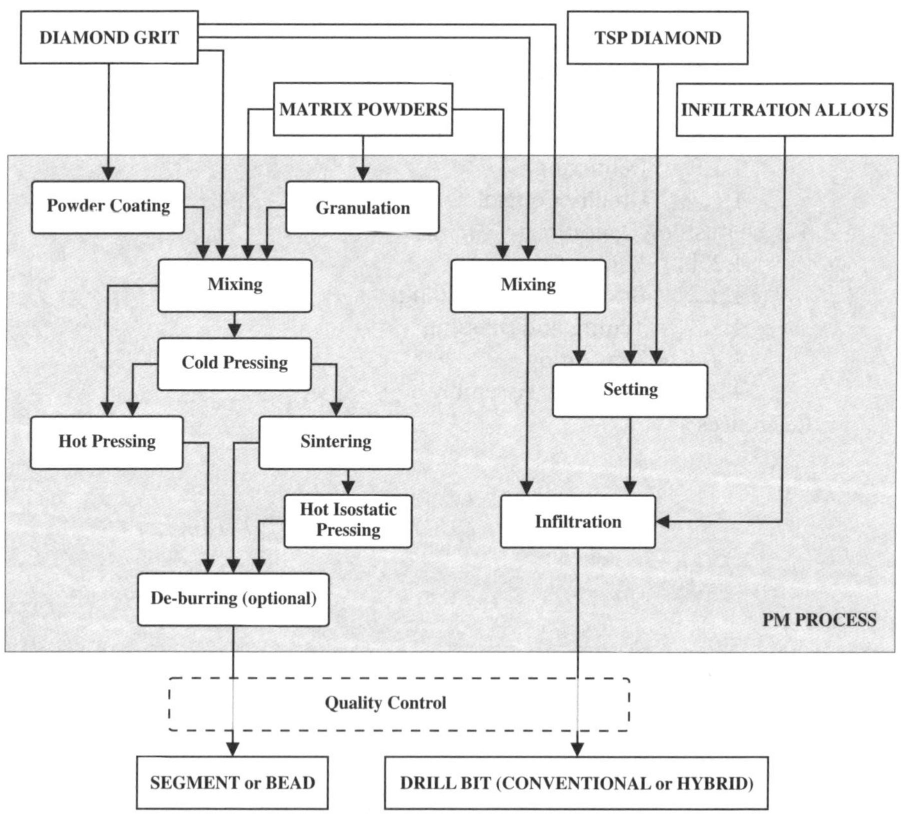

Figure 4.1 PM diamond tool production process.

4.1.1 Matrix Powder Preparation

Basically, the matrix powder preparation consists in mixing the selected component powders so as to achieve the pre-determined chemical composition, and particle shape and size distribution considering the final product application. This operation is often carried out using pendulum motion mixers.

Binding agents and lubricants, e.g. paraffin oil, monoethylene glycol, etc., are often added to the powder at this stage, in amounts up to 2 wt.%, so as to reduce dust and prevent segregation when the powder is subsequently handled or processed; but also to minimise wear of steel dies and to reduce oxides during subsequent cold and hot pressing operations, respectively.

4.1.1.1 Granulation

When the powder is to be cold compacted on a press working on a volumetric die filling principle, a further granulation treatment is necessary to obtain required flow and packing characteristics of the powder (see Fig. 4.2). Granulation may be carried out in a variety of ways [1–3] but, practically, two techniques have became widespread in the diamond tool industry. The older one, employed in the so-called 'Spartan' granulators, is based on high-speed mixing principle [1,3,4]. The latest

Figure 4.2 Matrix powder in as-granulated condition.

technique is based on a mechanical rolling (rounding) process, which additionally enables diamond particles to be 'ingranulated' in a single step process [5].

Irrespective of the processing route, organic binders, e.g. poly(butyl methacrylate), poly(alkyl methacrylate), polyvinyl butyral, paraffin wax, etc., dissolved in suitable solvents, are used to cement the powder particles together, thereby imparting desired mechanical strength to the granules. It is important that the binder has suitable thermal properties, which permit its complete removal from the material at the hot consolidation step. Otherwise, the segments have higher residual porosity, which may degrade their quality and create problems with brazing to the steel support.

4.1.2 Matrix–Diamond Mixture Preparation

The diamonds may either be used in as-received condition or coated with a suitable powder prior to mixing with the matrix powder [3–6]. Undoubtedly, the mixing process has a great effect on the quality of the final product. Non-uniform distribution of both matrix powder particles and diamond crystals will cause premature wear of the segment. Therefore when each diamond crystal is separately coated with the matrix powder, the formation of diamond clusters in the segment is practically eliminated and consequently better diamond distribution is guaranteed.

Another advantage of using powder-coated diamonds is longer life of the cold press steel dies due to efficient separation of hard abrasive particles from the compacting tool elements.

Finally, by proper selection of the coating composition aiming at increased resistance to wear, as compared to the surrounding matrix, it is possible to markedly increase the diamond particle protrusion, as shown schematically in Fig. 4.3, and thereby to improve the tool life and its free cutting characteristics [7].

The most widely used powder coating method consists in preparing a homogeneous slurry of the matrix powder with a solution of an agglomerating binder dissolved in a volatile organic solvent and spraying the slurry onto a bed of warm

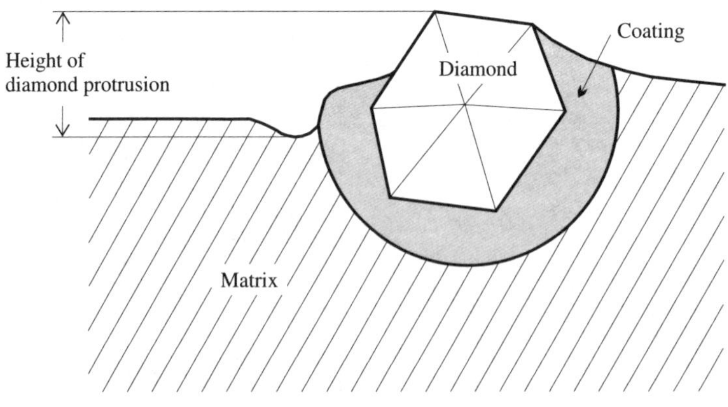

Figure 4.3 Positive effect of a wear-resistant coating on diamond particle protrusion.

air-fluidised diamond grits for a time until a coating of desired thickness has been deposited on each abrasive particle [4,6,8,9]. Glycerol, polyethylene glycol and liquid paraffin have been found as the most suitable binding agents, whereas ethyl alcohol, trichloro-ethylene and iso-propanol are typically used as solvents [4,6]. Experience has shown that the amount of the organic binder in the coating should not exceed 3 wt.%.

A final heating step may optionally be employed to make the diamond-containing granules stronger or even to eliminate porosity from the metallic coating [8].

4.1.3 Cold Pressing

Cold pressing is always applied as a pre-sintering operation and prior to hot pressing in the manufacture of saw segments having a layered structure (see Fig. 3.2).

Optionally, the cold pressing step can be utilised in fabrication of uniform tool components by the hot pressing route. Although additional equipment is required, by using green segments it is possible to increase productivity of the hot pressing process since, as exemplified in Fig. 4.4, the purpose-designed graphite mould yields more segments per hot pressing cycle than the conventional one filled with loose powder.

Typical cold pressing operations are performed in steel dies, at low to medium pressures, utilising the double-action pressing principle. There are two types of machinery used in the diamond tool industry. The conventional presses are fitted with either vibratory or screw powder feeders and precision scales used to weigh out the correct amount of the diamond–matrix mixture to fill the die. Alternatively, the machines may incorporate feed shoes operating on the volumetric filling principle. The gravimetric presses offer higher flexibility necessary to manufacture segments in smaller quantities [10]. By contrast, despite higher investment costs attributed in part to obligatory use of granulated powders, the volumetric equipment is the preferred option for mass production of small segments due to its 3–4 times greater throughput, longer life of steel dies and lower cost of other pressing consumables [11,12].

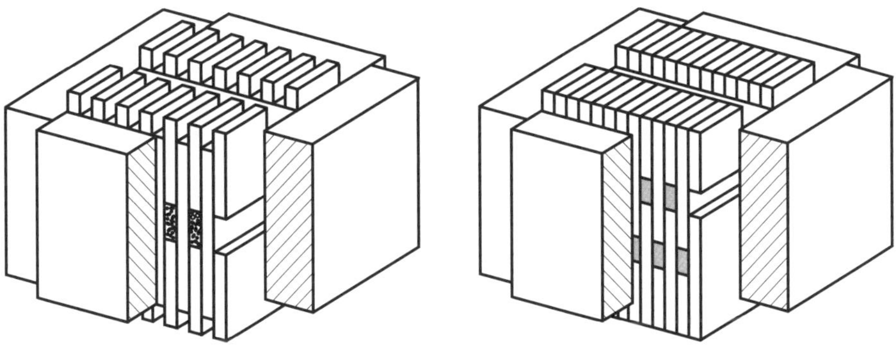

Figure 4.4 Graphite moulds designed to accommodate either loose powder mixture (left) or green segments (right).

4.1.4 Hot Pressing

The hot pressing process consists of a simultaneous application of heat and pressure so as to obtain a product nearly free from internal porosity. Compared to the conventional cold pressing/sintering PM route, hot pressing requires merely 2–3 min hold at markedly lower temperature, but under a compressive stress, to reach higher density level.

Due to the limited resistance of commercially available synthetic diamonds to elevated temperature, as well as to the growing demand for excellent mechanical properties of the matrix, the intrinsically rapid hot pressing technique has gained a widespread use in the production of diamond-impregnated tool components.

As shown in Fig. 4.5, hot pressing of diamond-impregnated segments, or beads, is generally realised in high-resistance graphite moulds by passing electrical current directly through the mould.

This method offers high efficiency in segment production and, at elevated temperatures, assists in protecting both the metal powder and diamond grit against oxidation. The protection is attributed to the formation of a CO/CO_2 reducing atmosphere inside the graphite mould which, in the old-type equipment shown in Fig. 4.6, is exposed to air.

The modern hot presses, exemplified in Fig. 4.7, are fitted with protective gas chambers wherein the moulds are heated in nitrogen and therefore their service life is markedly increased.

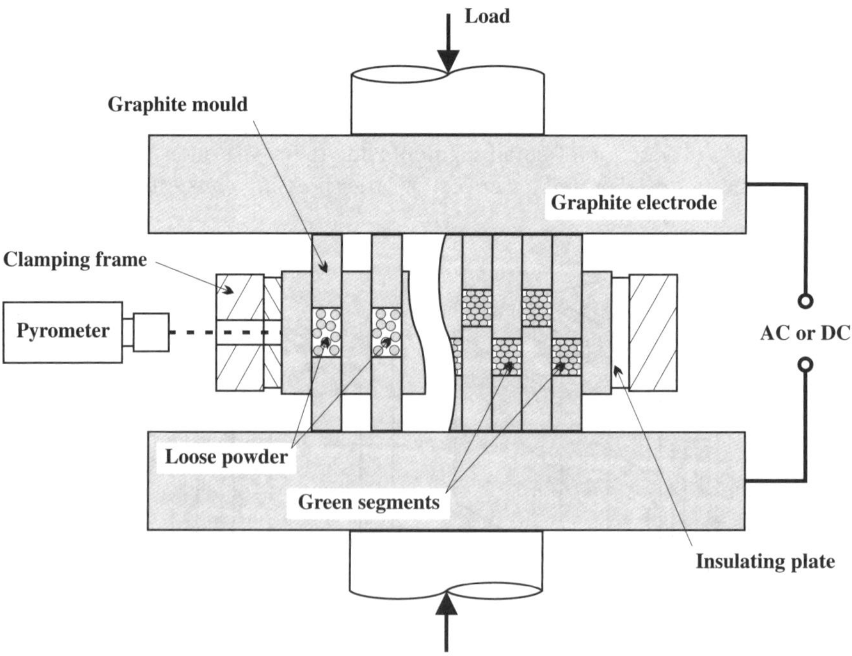

Figure 4.5 Schematic representation of the hot pressing process.

Figure 4.6 Hot pressing operation carried out in air.

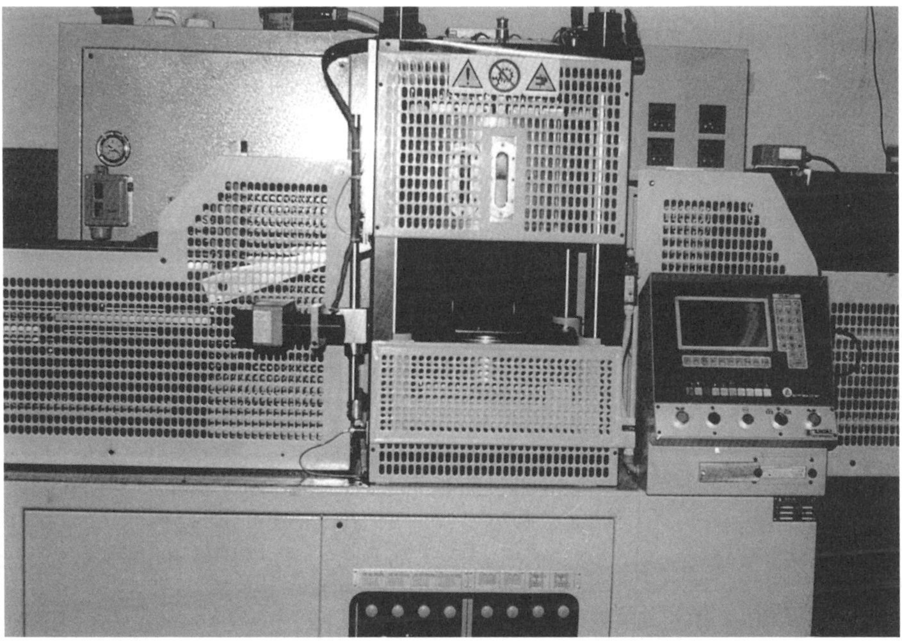

Figure 4.7 Hot press equipped with a protective atmosphere chamber.

The principle of direct resistance heating has some limitations with respect to the size and geometry of the tool components. Experience has demonstrated that the production of continuous rim sawblades and grinding wheels is slow and energy consuming due to large dimensions of these tools. Furthermore, difficulties are very often encountered in assuring uniform temperature distribution across the mould.

Therefore, more favoured equipment for this application is a furnace press wherein the graphite mould, or preferably steel die, is heated by radiation and convection [13,14]. To increase productivity, the furnace may consist of two separate stations for pre-heating/de-waxing and for final consolidation, respectively [14]. Modern furnace presses enable simultaneous processing of several steel dies, piled one on top of the other, in inert gas within less than 1 h and with several-fold lower power consumption as compared with the directly resistance-heated equipment.

4.1.5 Sintering

Sintering is an indispensable step in fabrication of numerous structural and tool PM components. Nonetheless, its application in the manufacture of diamond-impregnated tools is limited by restrictions on the composition, mechanical strength and dimensional accuracy of the final product. The common use of reducing hydrogen-containing atmospheres ensures an oxygen-free environment that is important to the protection of diamonds subjected to longer dwelling time at higher processing temperature, as compared with hot pressing, to obtain a fully dense product; but, such a treatment softens the matrix by removing oxide particles thus promoting recrystallisation and grain growth. Additionally, non-uniform density in the green segment, incorporating very fine matrix powders, inevitably leads to its distortion and lack of dimensional consistency after sintering.

The conventional cold pressing/sintering route has, however, been adapted on a broader scale to the production of wire saw beads [15–17], but can also be used for fabrication of continuous rim and occasionally segmental sawblades [18]. As opposed to other techniques, cost savings and higher rates of production achieved by the use of furnace sintering outweigh, in certain cases, the shortcomings of this operation.

4.1.6 Hot Isostatic Pressing

Hot isostatic pressing (HIP) consists in the application isostatic pressure produced by a hot inert gas to a suitably pre-consolidated powder mixture so as to obtain a fully dense material. The green parts are pre-sintered in a conventional sintering furnace to reach such a density when all pores are closed [17]. This makes the material suitable for direct HIP treatment. Alternatively, the pre-consolidated parts must be encapsulated in evacuated glass tubes prior to HIP as shown in Fig. 4.8. The semi-finished parts are loaded into the pressure vessel of a hot isostatic press, which is evacuated and purged several times with the pressurising gas, typically argon, before the HIP cycle begins [19].

The HIP units are capable of achieving pressures up to 200 MPa [20] and therefore allow marked reduction in the consolidation temperature as compared to the

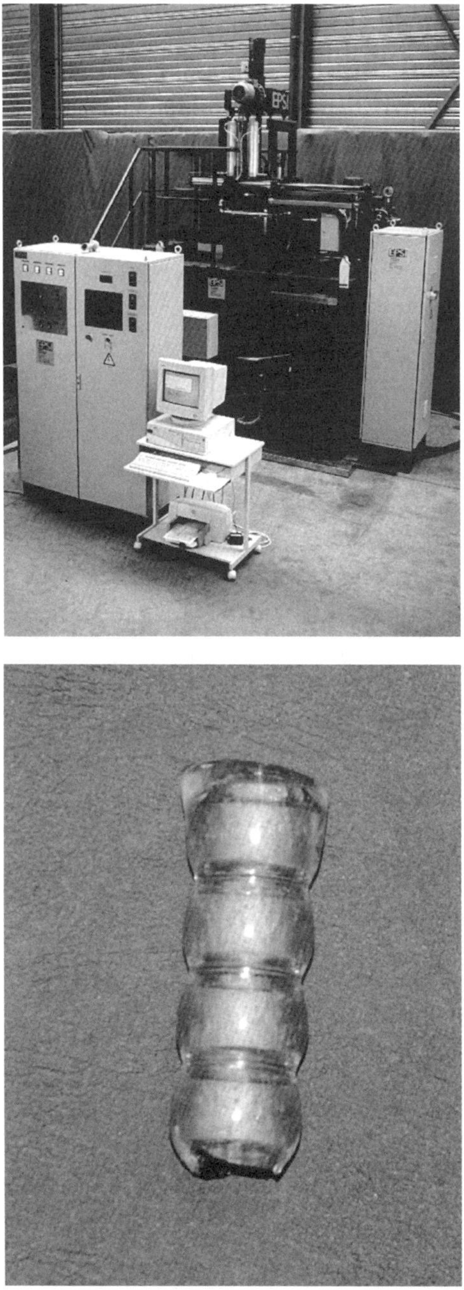

Figure 4.8 Medium-size hot isostatic press, used for production of diamond-impregnated beads and segments; and glass-encapsulated wire saw beads prior to final consolidation by HIP. (Courtesy of EPSI N.V.)

hot pressing technique. This opens up new technological opportunities because HIP can follow the hot pressing operation to eliminate residual porosity from segment compositions that require extremely high hot pressing temperatures, often exceeding 1100°C, with the risk of diamond degradation reduced to a minimum.

Over the last few years, the HIP technology has been gradually introduced into the fabrication of diamond tools with a marked trend to apply HIP in the mass production of wire saw beads [21].

4.1.7 Infiltration

Infiltration is ideally suited to make rotary surface-set diamond bits, TSP bits and TSP hybrid bits used in drilling in earth formations. As shown schematically in Fig. 4.9,

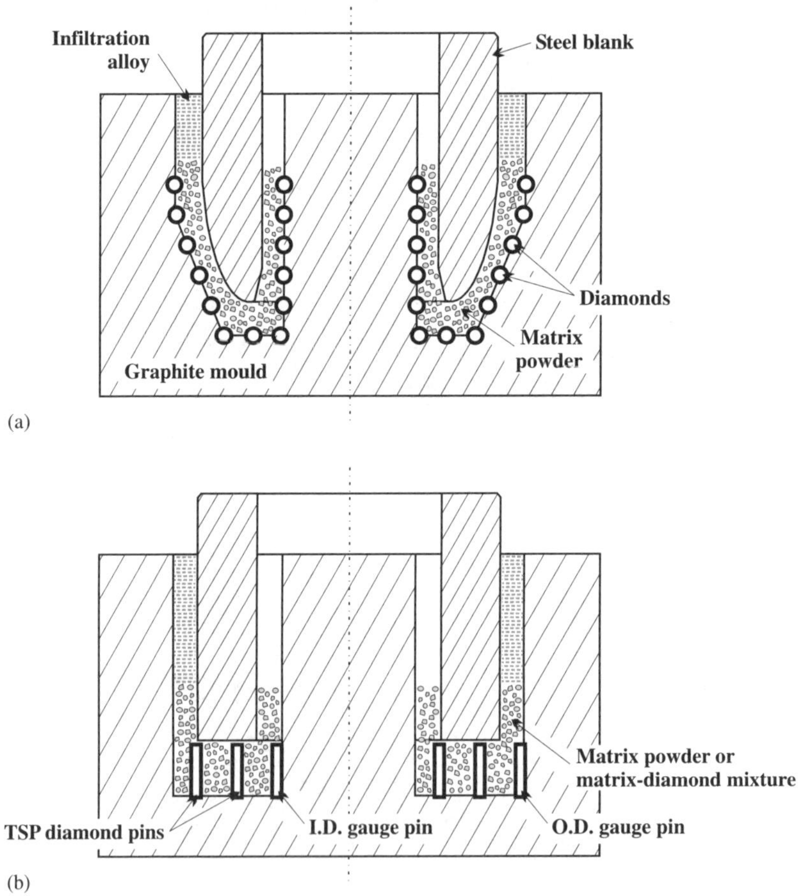

Figure 4.9 Sectional view of mould assemblies used for production of core head type (a) surface-set bits and (b) TSP (or TSP hybrid) bits.

the process consists in filling the interconnected pores of loose powders, placed in a graphite mould, with a liquid infiltration alloy having a lower melting point.

The infiltration process comprises a series of steps, beginning with an operation in which a single layer of coarse diamonds is positioned directly in the mould and arranged in a desired pattern to create the working face of the bit (see Fig. 4.9a). In the case of the TSP and TSP hybrid type bits (see Fig. 4.9b), thermally stable PCD rectangles are located across the mould base-plate with some extra pins placed in the gauge sections to reinforce the ID and OD portions of the bit [22]. Various displacement parts, which are removed after infiltration to provide internal drilling fluid passages, may also be mounted at this stage within the mould [23]. The diamonds are disposed in pre-machined sockets and secured with a conventional adhesive to prevent their dislocation during subsequent operations.

In the next step, a tubular steel blank is positioned within the mould and held in place by a suitable fixture. Then the matrix-forming powder, or matrix–diamond mixture, is introduced therein and compacted by tamping or vibration. The matrix powder usually comprises particles of tungsten carbide and/or eutectic tungsten carbide but metal powders, such as tungsten, iron, nickel, cobalt and copper, may also be used [24,25]. The particle shape, size and size distribution of the matrix powders should be carefully selected to attain required tap density of the material and also to avoid its shrinkage during high-temperature processing. Generally, it is undesirable to use very fine matrix powders since they sinter readily leading to inconsistent composition and dimensions of the final product. Small amounts of fine powders may, however, be used to fill the interstices between coarse particles without encountering any sintering problems. To ease mould loading, the powdered material is sometimes applied in a liquid hydrocarbon carrier [23], which vaporises when heated.

Once the powder has been placed in the mould, in such a way that it surrounds a certain portion of the steel blank, a suitable infiltration alloy, preferably in the form of coarse shot, pellets or pre-cut chunks, is positioned adjacent to the matrix powder. The amount of infiltration alloy should be carefully calculated so that there is a slight excess of liquid to be machined off after solidification. Various copper-base materials containing zinc, nickel, tin and manganese as the major constituents are readily used as infiltration alloys [24]; but small amounts of other elements, such as iron, molybdenum, silicon, led, silver, cadmium, phosphorus and chromium [24–26], can also be added to enhance wetting, lower the infiltration temperature and reduce dissolution of the solid matrix components in the melt.

The whole mould assembly is placed in a furnace and heated so that the infiltration alloy melts and is drawn, by capillary action, into the powder mass bonding it to the steel blank. Depending on the melting range and viscosity of the infiltration alloy, the process is carried out at temperatures typically ranging between 950 and 1200°C [27], preferably in a reducing atmosphere [26] or vacuum [25–27]. The controlled atmosphere may not be needed if a suitable low-melting-point flux is applied [27].

After infiltration, the mould is cooled down. Then the bit is removed from the mould and, if the steel blank has not been suitably prepared prior to infiltration, its projecting end is welded to a threaded upper body to fit a drill string.

4.1.8 Other Consolidation Routes

In addition to the above-mentioned processes, there are other ways in which a mixture of diamond grit and bonding powders may be consolidated to form a cutting tool, e.g. by brazing [28,29], laser cladding [30,31] or extrusion [32]. Of these, the first two techniques have gained some industrial significance and deserve a concise description here.

4.1.8.1 Brazing

This technology consists in applying a thin uniform coating of braze-diamond slurry on the working face of a suitably shaped tool body and heating the part thus prepared in vacuum to produce a single layer of diamond grits firmly bonded to the substrate [28]. Alternatively, the diamonds can be set on the tool in a regular manner [29], which enables the spacing between individual grits to be optimised and prevents the formation of either multiple-coated or diamond-depleted areas. In this case, the steel carrier is first covered with an array of tiny glue drops by means of an automatic glue dispenser [29]. Diamond grits are then sprinkled and glued to the surface. Finally, a suspension of suitable braze powder is sprayed over the diamond prior to placing the part in the furnace.

Both so-called 'active' brazes and nickel-base brazes are suitable for this application [28]. The active brazes are relatively soft silver-base alloys containing titanium as the active element, which improves the alloys' wetting characteristics and aid in chemical bonding of the diamonds. The nickel-base alloys are mechanically stronger and contain a combination of chromium and silicon, which elements play an equally important role as titanium does in the active brazes. The optimum brazing temperatures range from 900–950°C up to 1000–1050°C for the active and nickel-base brazes, respectively.

The main advantages of tools produced by brazing are: small consumption of diamond grits; perfect control over amount and distribution of cutting edges; and excellent grit protrusion. As a result, the 'brazed' tools offer a good combination of high cutting rate, low power consumption and extended tool life.

4.1.8.2 Laser cladding

A technique that is coming increasingly into use in powder metallurgy is laser sintering. A modification of this technique has also been implemented into the diamond industry to produce small circular sawblades for sawing stone, concrete and asphalt [30,31].

The laser cladding process involves a direct deposition of diamond-impregnated material onto the rim of a low-carbon steel centre. The diamond grit and matrix powder are supplied in a controlled manner to a mixing unit and subsequently injected through a nozzle into the focal point of a laser beam, wherein the matrix powder is melted in an inert atmosphere and fused to the substrate. The process takes place adjacent to the rim of a slowly rotating steel centre that is securely clamped between two chilled moulds. The moulds project beyond the steel centre to form a cavity, which determines the thickness and shape of the cutting rim. By simultaneously controlling the matrix-diamond mixture supply and feed rate of the substrate, it is possible to deposit either segmental or continuous cutting layer of a suitable height.

Rapid solidification of the metallic melt, results in a fine-grained structure and excellent mechanical properties of the matrix material. Tin bronzes appear to be ideally suited for laser cladding and are exclusively used as the base alloy to which titanium is added to enhance wetting and to reinforce the hold on the diamonds by chemical means [31]. Moderate additions of tungsten carbide powder may also be applied to increase the resistance to wear of the matrix [31].

4.1.9 Deburring

Some diamond-impregnated-tool components, e.g. sawblade segments, grinding segments, etc., require cleaning and removing edge residuals after consolidation. This is carried out during the deburring operation, which is usually performed by means of tumbling the segments with coarse alumina or silicon carbide grit.

4.1.10 Quality Control

The quality control of diamond-impregnated tools, or tool components, is frequently limited to a hardness test [33]. The Rockwell B test is the most widespread technique due to its simplicity and inexpensiveness. A properly densified matrix-diamond mixture acquires a narrow hardness range which, to a great extent, is affected by the matrix composition. On the other hand, if the structure of the segment deviates substantially in any respect, or if the densification is incomplete, the hardness does not fall within the specified range. Incompletely densified materials usually have extremely low toughness, which results in poor wear resistance and diamond retention properties of the matrix. Therefore, if there is any doubt about the hardness readings, the evaluation of the final product density becomes another important quality check [33]. To secure accurate density values, the measurements must be performed by means of the water-displacement technique [34].

In the case of infiltrated drill bits, visual examination is routinely used to assure correct diamond setting and protrusion but non-destructive techniques, such as radiography, may also be utilised to detect internal flaws and discontinuities in the bit crown [35].

4.2 Finishing Operations

In the vast majority of situations, the end product of the PM processing cannot be directly applied as a tool and must be subjected to further finishing operations. These are mainly connected with attaching the diamond-impregnated segments, or beads, onto a suitable steel carrier that is also prepared to cope with the forces applied during use.

4.2.1 Radius Grinding

A great range of different sawblade diameters would require the toolmaker to keep in stock all the shapes of segments and a large number of various graphite moulds used to manufacture these segments. To solve this problem, a single segment type having a diamond-free base is often produced to fit any steel centre in a pre-set range of diameters after the required radius has been ground on the base.

The radius grinding operation may also be needed to prepare segments for laser welding. Modern laser welding equipment uses a focused laser beam 0.1 mm wide. Therefore the radius on the base of the segment has to be executed with great accuracy in order to ensure that the segment perfectly fits the steel centre.

These days, highly sophisticated equipment, capable of sorting the segments according to size, is being designed for mass production use [36] but the older-type swinging arm grinders and manual grinding can also be used.

4.2.2 Brazing/Laser Welding

Diamond-impregnated segments need to be attached to a suitably shaped steel support to obtain a sawblade, core drill or grinding wheel. In general, either brazing or laser welding may be used at this stage. Brazing is well established in the manufacture of tools for wet processing of natural stone, which can be re-tipped; whereas laser welding is mainly used in the mass production of small diameter, dry cutting circular blades. In the latter case, the heat generated during sawing softens conventional braze joints and hence there is a risk that a whole segment breaks off the steel centre during a high rotational speed dry cutting operation. Since the bending strength of a laser-welded seam may achieve 1800 MPa, compared with 350–600 MPa for the brazed joint [37], laser welding practically eliminates the possibility of segment detachment.

The brazing operation may be carried out either on a dedicated brazing machine fitted with high-frequency induction heater controlled by an optical pyrometer [38] or manually using a gas torch. Prior to the application of temperature, fluxed segments and a braze foil are positioned on the edge of a fluxed tool body and attached by mechanical means or magnets.

It is essential that the heat and thermal stresses generated during brazing do not weaken and/or deform the tool shank. It may become critical in brazing segments to circular sawblades, where temperatures above 700°C should generally be avoided and brazing time must be kept to a minimum. In this case, the best solution is to use a microprocessor-controlled brazing machine equipped with a water-cooled clamp that protects the slot-base section of the steel centre from overheating.

Laser welding imposes restrictions in selection of the steel centre and segment matrix material. Low-carbon steels of low hardenability are most frequently used to avoid formation of brittle martensite in the heat-affected zone due to rapid cooling after the laser beam has passed. For that reason, the segment must also have a diamond-free base made of a suitable material to yield a strong flawless welded joint with the steel.

An important advantage of laser welding is its high productivity with operator intervention reduced to a minimum [37]. The capital expenditure is high, but may fully be compensated for by very high volume production, e.g. in the low-price DIY tool sector.

4.2.3 Truing and Dressing

The objective of the truing operation is to render the segments concentric with the bore of the tool body, to clean their sides and reduce lateral run-out. The subsequent dressing operation is performed so as to remove the matrix from around the diamond particles, to produce sufficient protrusion and allow efficient cutting from the outset. Contrary to truing, dressing is carried out in the down-grinding mode, i.e. the blade and grinding wheel turn in the same direction which exerts less impact on diamond particles [39].

4.2.4 Tensioning

Tensioning is an established practice in the manufacture of diamond sawblades since the segments themselves, even when perfectly tailored to the application requirements, do not guarantee the proper action of the whole tool [40,41].

The circular steel centre accumulates stresses during fabrication. Since they are never perfectly symmetrical, the sawblade is incapable of spinning without wobble. Therefore, the steel centre must initially be subjected to an operation of neutralising the unequal stresses to cause it to lie flat. Skilful 'levelling', as it is generally termed, is then followed by the actual tensioning which consists in the addition of extra stress to stretch the centre section, which counteracts the centrifugal force tending to elongate the rim section of the rotating blade.

Special precautions must be taken during material selection and tensioning of sawblades intended for application on multi-blade tile sawing machines [42]. They operate with a large number of sawblades ganged closely together on a single spindle and therefore, the failure of a single tool is immediately transmitted to the others and involves carrying out complicated operations to recondition the whole set of sawblades.

Except the final checks of the blade's lateral run-out and stress condition, which have been fully mechanised and computerised [43–45], the steel centre mechanical servicing remains an arduous effort consisting of rolling and hammering by a skilful and experienced saw smith.

4.2.5 Wire Saw Assembly

At the PM stage, the production of wire saw beads yields a semi-finished product comprising a fully consolidated diamond-impregnated ring firmly secured on a solid cylindrical core, which subsequently require to be bored through and threaded [16]. The beads and spacers are then alternatively mounted at regular intervals on a steel rope by means of simple assembly equipment [46] or automatic machines [47]. It is essential for the tool performance that the diamond beads are tightly held in place and do not rotate around the rope. The recent tendency is to use injection-moulded plastic or vulcanised rubber to separate the individual beads and to provide a water-tight protection of the rope against abrasion by aggressive slurries [48]. The older type of diamond wire, with helical spring spacers and stopping rings [49], which can easily be assembled or repaired in the field, may still be employed for sawing marble and other non-abrasive materials.

To achieve rotation of the wire in the cut, so as to produce a uniform wear over the whole working surface of the beads, the wire saw is suitably pre-twisting around its longitudinal axis before the continuous loop is assembled by means of a clamping pipe or threaded connecting sleeves.

References

1. Moriya, N., Granulating apparatus. US Patent 4,655,701 (April 7, 1987).
2. Weber, G., Granulating: a new process for diamond tool producers. In Proceedings of International Workshop on Diamond Tool Production, Turin, Italy, November 8–10, 1999, pp. 73–82.
3. Baert, M., Coating of diamonds and granulation of metal powders. In Proceedings of Seminar on PM Diamond Tools, Lausanne, Switzerland, November 2–3, 1995, pp. 24–40.
4. Van Doorslaer, T., Coating of abrasive grains and the granulation of metal powders. In Proceedings of International Workshop on Diamond Tool Production, Turin, Italy, November 8–10, 1999, pp. 83–88.
5. Burckhardt, S., New technique for granulating diamond and metal powders. Industrial Diamond Review, 1997, 57(4), 121–122.
6. Kimura, K., Method for forming metal-coated abrasive grain granules. US Patent 4,770,907 (September 13, 1988).
7. Cai, O., Encapsulated diamond: remarks and technical notes. Diamante Applicazioni & Tecnologia, 1996, 2(8), 62–72.
8. McEachron, R.W., Clark, T.J., Matarrese, R.R., Sinigaglia, S., Coated abrasive particles. European Patent Application 0533444A2 (September 15,1992).
9. Matarrese, R.R., Dual-coated diamond pellets and saw blade segments made therewith. US Patent 5,143,523 (September 1, 1992).
10. Weber, K., Modern diamond segment production on cold presses and hot presses. In Proceedings of Seminar on PM Diamond Tools, Lausanne, Switzerland, November 2–3, 1995, 48–50.
11. Weber, G., Burckhardt, S., Economic production of diamond segments. Industrial Diamond Review, 1998, 58(4), 111–112.
12. Anon., New high-performance cold press from Dr. Fritsch. Industrial Diamond Review, 2003, 63(4), 25.
13. Gartner, B.J., Development of vacuum sinter press for production of diamond tools. Industrial Heating, 1998, 65(10), 67–68.
14. Morelli, E., Oppici, S., New range of machines for the production of grinding wheels. Industrial Diamond Review, 2001, 61(2), 102–103.
15. Cai, O., Sintering without pressure, or free sintering. Diamante Applicazioni & Tecnologia, 1997, 3(11), 106–122.
16. Morelli, E., Developments in diamond wire bead manufacture. Industrial Diamond Review, 2000, 60(3), 223–224.
17. Bonneau, M., Moltenni, M., Wire manufacturing and free sintering with NEXT. Industrial Diamond Review, 2002, 62(4), 263–265.
18. Sakarcan, M., Method of manufacturing a segmented diamond blade. US Patent 5,471,970 (December 5, 1995).
19. Bose, A., Eisen, W.B., Hot Consolidation of Powders and Particulates. MPIF, Princeton, NJ, 2003.

20. EPSI isostatic presses. Engineered Pressure Systems Incorporated, 2003, Product information brochure EV-2003.
21. Davis, P.R., The future of diamond abrasives in stone processing. Industrial Diamond Review, 2001, 61(3), 159–167.
22. Sheppard, N., Dolly, B., Hybrid bit success for SYNDAX3 pins. Industrial Diamond Review, 1993, 53(6), 309–311.
23. Smith, R.H., Cooley, C.H., Rotary drill bit with abrasion and erosion resistant facing. US Patent 4,884,477 (December 5, 1989).
24. Matrix powders for diamond tools. Kennametal Inc. catalogue, 1989.
25. Naidich, Y., Diamond grit bits of increasing working life with strong chemical fastening of cutting elements in matrix. In Proceedings of PM 2001 Congress and Exhibition, Nice, France, October 22–24, 2001, pp. 434–438.
26. Griffin, N.D., Manufacture of rotary drill bits. US Patent 4,669,522 (June 2, 1987).
27. Horton, R.M., Anthon, R.A., Low melting point copper–manganese–zinc alloy for infiltration binder in matrix body rock drill bits. US Patent 5,000,273 (March 19, 1991).
28. Trenker, A., Seidemann, H., High-vacuum brazing of diamond tools. Industrial Diamond Review, 2002, 62(1), 49–51.
29. Boretius, M., Burkhard, G., Vacuum brazing of hard and super hard materials. In Diamond Tooling Proceedings of Euro PM 2002, Lausanne, Switzerland, October 7–9, 2002, pp. 90–94.
30. Grüneis, H., Grüneis, T., Sintering and brazing all in one. Industrial Diamond Review, 1998, 58(2), 45–47.
31. Gottschling, S., Luber, A., Grüneis, T., Grüneis, H., New blown powder laser cladding process for the production of cutting tools. Diamante Applicazioni & Tecnologia, 1999, 5(18), 99–102.
32. Samvelion, R.V., Manoukion, N.V., Extrusion for diamond tool production. In Proceedings of PM Diamond Tools Seminar, Lausanne, Switzerland, November 2–3, 1995, pp. 65–72.
33. Konstanty, J., Cobalt as a Matrix in Diamond Impregnated Tools for Stone Sawing Applications. AGH Uczelniane Wydawnictwa Naukowo-Dydaktyczne, 2nd Edition, Krakow, 2003.
34. Standard test methods for metal powders and powder metallurgy products. MPIF Standard 42 (issued 1980). MPIF, Princeton, NJ, 1996.
35. Weitang, L., Diamond bit manufacturing technology. In Diamond Drilling Handbook, edited by Guangzhi, L., Geological Publishing House, Beijing, China, 1992, pp. 272–328.
36. Burckhardt, S., Urtel, E.V., Fully automatic grinding of radii on diamond segments. Industrial Diamond Review, 2000, 60(2), 135–136.
37. Weber, G., Laser welding of diamond tools. Industrial Diamond Review, 1991, 51(3), 126–128.
38. Davies, A., Induction brazing for diamond toolmakers. Industrial Diamond Review, 1992, 52(6), 325–327.
39. Wright, D.N., Ford, G., The truing and dressing of diamond sawblades. De Beers Industrial Diamond Division internal paper.
40. Anon., Tensioning of a diamond blade. J. Chaland & Fils.
41. Pratt, W.R., Tension of diamond sawblades. In Proceedings of 2nd DWMIT Technical Symposium Sawing and Grinding with Diamond Wheels, Chicago, IL, March 7, 1973, pp. S9/1—S9/4.
42. Jennings, M., Multi-blade teach-ins at Chaland. Industrial Diamond Review, 1993, 53(1), 6–7.
43. Büttner, A., Mummenhoff, H., Testing the stress in diamond circular sawblades for sawing natural stones and concrete. Industrial Diamond Review, 1973, 33(5), 376–379.
44. Huelmann, E., Kohl, H.W., Optimum quality control of sawblade centres. Industrial Diamond Review, 1991, 51(6), 278–279.
45. Weber, G., New software for automatic tensioning of sawblades. Industrial Diamond Review, 2004, 64(2), 55–56.
46. Anon., Diamond tools for the stone industry. Diamant Boart, S.A, 1995.
47. Anon., Automatic threading and unthreading of diamond wires. Industrial Diamond Review, 2004, 64(1), 63–64.
48. Diamond wires. Co.Fi.Plast, Product information brochure.
49. Diamond wire. Maco s.r.l., Product information brochure.

CHAPTER 5

Microstructure of the Matrix

The microstructure of a material influences its properties such as the mechanical strength, hardness, ductility, resistance to wear, etc. It is therefore of key importance to understand the relationship between the properties of raw powders, processing conditions and microstructure of the consolidated product.

5.1 Processing to Near-full Density

This is obviously the matter of the utmost concern that the matrix, wherein the diamond grits are embedded, has been processed to virtually theoretical density prior to application. It is well documented that the finer the powder the more prone it is to densification [1–12].

Typical hot pressing curves for different cobalt powder grades are presented in Fig. 5.1.

It is generally recommended that the pressure is restricted to around 35 MPa, which is the safe limit for most mould graphite grades.

Additions to cobalt of pre-alloyed bronze series powders, elemental tin or copper, used separately or in combination, decrease the hot pressing temperature and/or pressure, whereas the reverse is observed in the case of tungsten and tungsten carbide powders.

The situation may change when powders are consolidated by the conventional cold pressing/sintering route. Fig. 5.2 shows the as-sintered densities of three commercial cobalt powders obtained from different production routes.

As would be expected the densification of green compacts is facilitated when sintering is carried out in hydrogen. It is important to note, however, that certain cobalt powders produced by the hydrometallurgical process, such as the Umicore Ultrafine grade, which are virtually free from oxygen (see Section 3.2.1), may also be sintered in vacuum to achieve near-full density at reasonably low temperature. On the other hand, difficulties are encountered in sintering powders containing oxygen combined as stable oxides, e.g. the Eurotungstene CoC grade, which are non-reducible with hydrogen under the sintering conditions.

The pre-alloyed cobalt alternatives (see Table 3.3) are generally characterised by an excellent compressibility at elevated temperatures. As documented in Fig. 5.3, for most of them the near-full density can be attained at between 650 and 800°C.

Certain matrix powders characterised in Fig. 5.3 shall, however, pose densification problems when subjected to the pressure-less sintering process, as illustrated in Fig. 5.4.

From Fig. 5.4 it is apparent that the Next 200, Next 900, and Cobalite CNF powders are well suited to densification by furnace sintering, whereas it proves difficult to eliminate porosity in Cobalite 601 and Next 300 which, as in the latter case, may retain intolerable amount of pores even after sintering at 1100°C.

In addition to the capacity of the powder to densification by sintering, it is equally important to what extent the matrix interacts with the diamond during prolonged exposure to higher temperature as compared with consolidation by hot pressing.

The high-grade diamond grits shown in Fig. 5.5, which have been extracted from sintered segments, exemplify the large difference in magnitude of surface attack exerted by two different iron-base matrices.

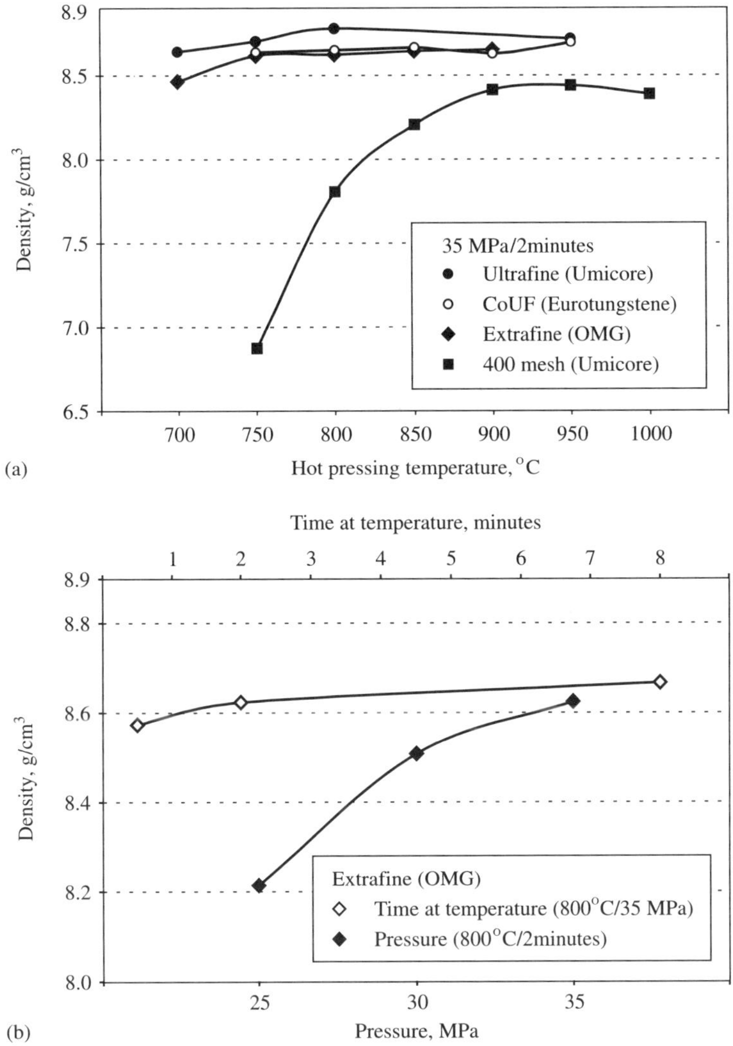

Figure 5.1 *Hot pressing curves showing dependence of final density on (a) powder type and temperature (compiled from Refs. [1,5,12]), (b) time at temperature and pressure [1].*

Diamonds embedded in Cobalite CNF retain their original colour and shape, whereas Cobalite 601 powder, which requires very high sintering temperature of 1100°C to be properly consolidated (see Fig. 5.4), causes massive degradation of diamond crystals, through their dissolution in austenite, thus disqualifying the product for application.

Experience has demonstrated that oxygen contained in the powders may also have a harmful effect on the surface and edge quality of the diamond grits especially when the sintering process is carried out in vacuum [19].

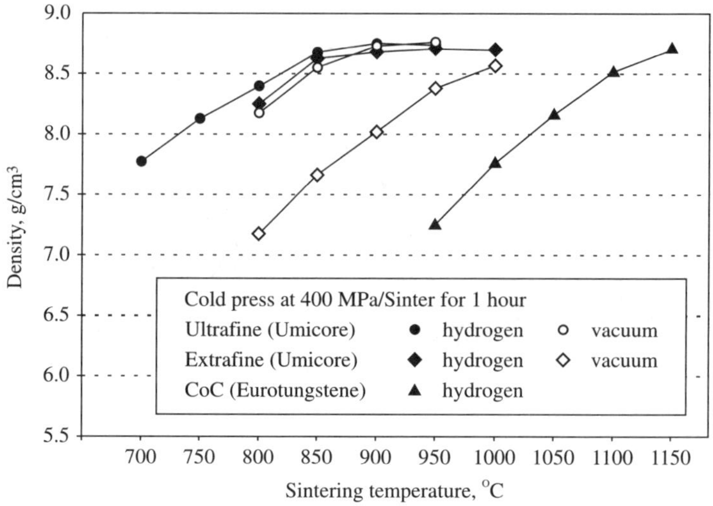

Figure 5.2 Sintering curves for selected cobalt powders.

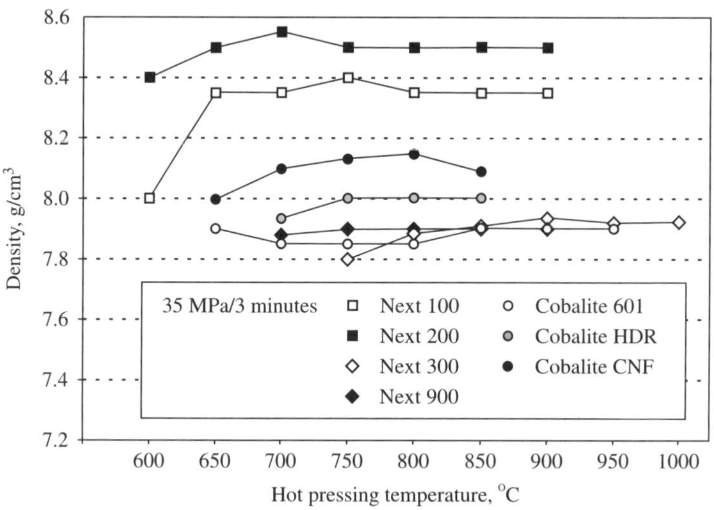

Figure 5.3 Hot pressing curves for selected commercial matrix powders used as replacements for cobalt (compiled from Refs. [8,13–18]).

Another disadvantage of the conventional cold pressing/sintering route is a tendency to form microcracks in the matrix at the cooling-down stage after sintering. They usually originate from sharp diamond edges and can be detected as surface defects during the final inspection of wire saw beads [20] or in a fractured material, as shown in Fig. 5.6.

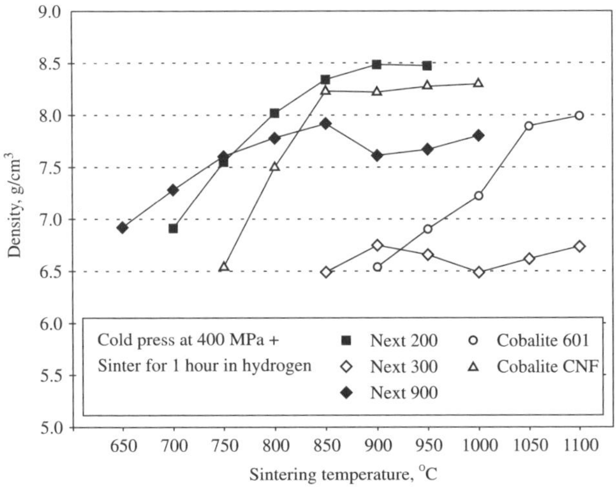

Figure 5.4 Sintering curves for selected cobalt substitutes.

(a)

(b)

Figure 5.5 Diamond grits extracted from segments sintered for 1 h in hydrogen to near-full density: (a) no evidence of chemical attack after sintering in Cobalite CNF at 850°C; (b) complete degradation of crystals sintered in Cobalite 601 at 1100°C.

Figure 5.6 Fracture surface of sintered cobalt matrix showing extensive cracking within diamond pullout sites.

5.2 Grain Size

The possibility of increasing the mechanical strength of a metal by reducing the grain size has long been observed and utilised in steels and other common metallic materials. The grain size strengthening occurs as a result of restrictions imposed on the movement of dislocations by the greater grain boundary surface area, and is quantified by means of the Hall–Petch equation

$$\sigma_y = \sigma_o + K d_g^{-1/2} \tag{5.1}$$

where σ_y is the yield strength, d_g the average diameter of the grains and σ_o and K are constants for the metal.

In diamond-impregnated tool components the grain size of the matrix can be affected by varying the properties of the starting powders and conditions for their consolidation, as exemplified in Fig. 5.7. The most important powder characteristics include its chemical and phase composition, chemical homogeneity, mutual solubility of component elements, amount of impurities such as oxygen and sulphur, mean particle size and particle size distribution.

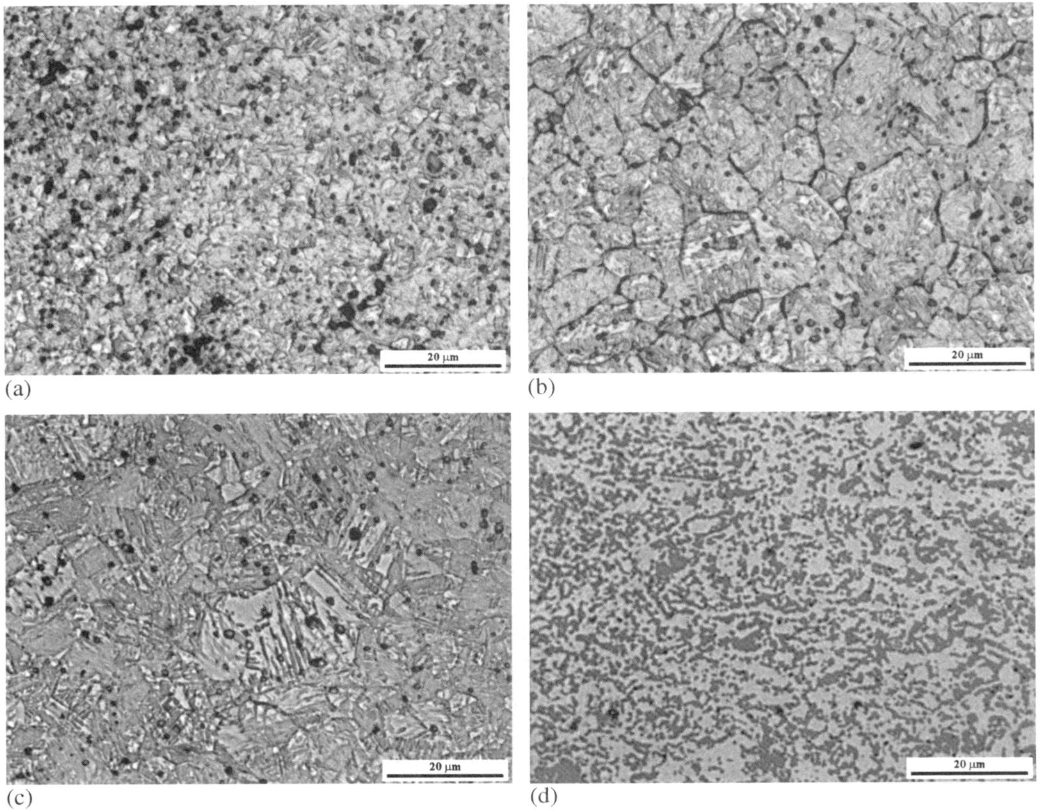

Figure 5.7 *Microstructures of single- and multi-component matrices after consolidation to near-full density: (a) Co Extrafine (Umicore) hot pressed for 2 min at 950°C and 35 MPa; (b) Co Ultrafine (Umicore) hot pressed for 2 min at 950°C and 35 MPa; (c) Co Extrafine (Umicore) sintered for 1 h at 950°C in hydrogen; (d) Next 200 sintered for 1 h at 900°C in hydrogen.*

Except in the infiltrated tools (see Section 4.1.7), very fine powders are the preferred choice for most applications. Such powders usually yield a matrix characterised by a fine-grained microstructure after consolidation by hot pressing (Fig. 5.7a). During the pressure-assisted densification, a combination of low processing temperature, short time at temperature and presence of finely dispersed oxides may effectively impede grain growth in the majority of the matrix compositions.

It is well known that oxides retard grain growth by exerting a pinning action on the grain boundaries. This mechanism may prove ineffective when the grain growth is stimulated by a liquid phase penetrating the grain boundaries, as has been observed in cobalt contaminated with sulphur and consolidated at a too high temperature (Fig. 5.7b) [12]. It is likely that the liquid $Co–Co_4S_3$ eutectic, which may locally appear above 877°C [21], penetrates grain boundaries and enables their rapid migration, notwithstanding the presence of a dispersed oxide phase [22].

In contrast to hot pressing, the consolidation of matrix powders by means of the cold pressing/sintering process generally favours grain growth due to the longer processing at higher temperature in a reducing gas. The common sintering atmospheres enable the reduction of unstable oxides and thus grain growth frequently concurs with the attainment of near-full density when the residual oxides and pores cannot efficiently block the movement of grain boundaries (Fig. 5.7c).

Pores are an inherent part of virtually all sintered materials. There is a thermodynamically justified affinity between the pores and grain boundaries that gives a higher chance that a pore will remain attached to a grain boundary thus contributing to deceleration of grain growth [23]. Examples are available [12] that around 5% porosity in the form of finely dispersed voids is capable to impede grain growth in cobalt.

If porosity is the controlling factor, the expected grain size can be connected with the amount and size of the pores by the general relationship [24]

$$d_g \propto \frac{d_p}{V_p} \tag{5.2}$$

where the average sizes of the grains and pores are denoted as d_g and d_p, respectively, and V_p is the volume fraction of the pores.

From relation (5.2) it is evident that the inhibiting effect of pores on grain growth decreases rapidly when the sintered single-phase material approaches its theoretical density. In that case the fine-grained structure cannot be preserved except in situations when other grain growth inhibitors are present in the material at the sintering temperature. These include finely dispersed stable oxides, which can intentionally be introduced to the material in a suitable amount at the powder production stage [18], and the presence of a mixed-phase solid structure throughout the isothermal sintering stage (Fig. 5.7d), that occurs in many dual- and multiple-component systems [25].

5.3 Recovery and Recrystallisation

The compacting pressure applied to the powder confined in a die, or mould, has to be above the yield strength at the temperature applied. This leads to the strengthening of the material by increasing the density of lattice defects, such as dislocations, vacancies, interstitial atoms, etc. At the hot consolidation step, recovery occurs at slightly elevated temperatures and enables some stress relief in the material due to the reduction in number of point defects and rearrangement of dislocations. Recrystallisation occurs at higher temperatures, typically above 0.4 homologous temperature, and eliminates the strain hardening by the nucleation and growth of new grains containing few dislocations. Because the density of lattice defects is greatly reduced, the recrystallised metal has a lower strength but a higher ductility.

The temperature at which a metal undergoes complete recrystallisation depends on a number of factors. The most important are the melting temperature, amount of cold work, time of annealing, and presence of oxides or other finely dispersed phases.

The influence of oxides on recrystallisation is complex. It has been proved experimentally that oxide particles randomly dispersed throughout a metal provide preferred nucleation sites by distorting its crystal lattice. However, the finer the particles the less faulted the structure around them and, consequently, the more difficult the nucleation of new grains [26]. Equally important is the oxide inter-particle spacing. It has been established that the oxide phase can effectively retard nucleation when the critical recrystallisation nucleus is bigger than the oxide inter-particle distance [26]. Besides, the impeding action of finely dispersed oxides facilitates better stress-relief through prolonged recovery thereby lowering the driving force for recrystallisation.

As documented by the X-ray back reflection patterns in Fig. 5.8, cobalt containing moderate amounts of oxygen shows little evidence of recrystallisation even after hot pressing at virtually 0.7 homologous temperature (see Fig. 5.8c); whereas advanced recovery and recrystallisation are evident in a low-oxygen material (see Figs. 5.8a and 5.8b).

For the coarsest 400 mesh powder hot pressing at around 950°C is common. Despite lower tendency to oxidation the material retains typically up to 5% of residual porosity, which acts in the similar way as fine oxide particles, and thus resists recrystallisation during the high temperature processing (see Fig. 5.8d). Interestingly, earlier studies of this 400 mesh grade, cold compacted and furnace sintered in either vacuum or hydrogen atmosphere, have indicated incomplete recrystallisation even after 1 h hold at 1300°C [27]. The powder contains relatively high amounts of calcium and silicon (see Table 3.2). These elements are well known to form extremely stable oxides [28], which remain intact up to at least 1300°C to hinder recrystallisation and grain growth.

Generally, the higher the oxygen content of the powder the stronger the effect of the oxide phase on impeding nucleation of new grains and grain size stability with temperature. Consequently, the material resists recrystallisation and retains high hardness far beyond its complete densification temperature. Such a reduced sensitivity of the matrix powder to temperature contributes to the formation of an elongated plateau on the hardness vs. consolidation temperature curve and is recognised in the tool-making practice as a feature that guarantees consistency in the segment's quality despite unexpected fluctuations in the pre-set processing temperature.

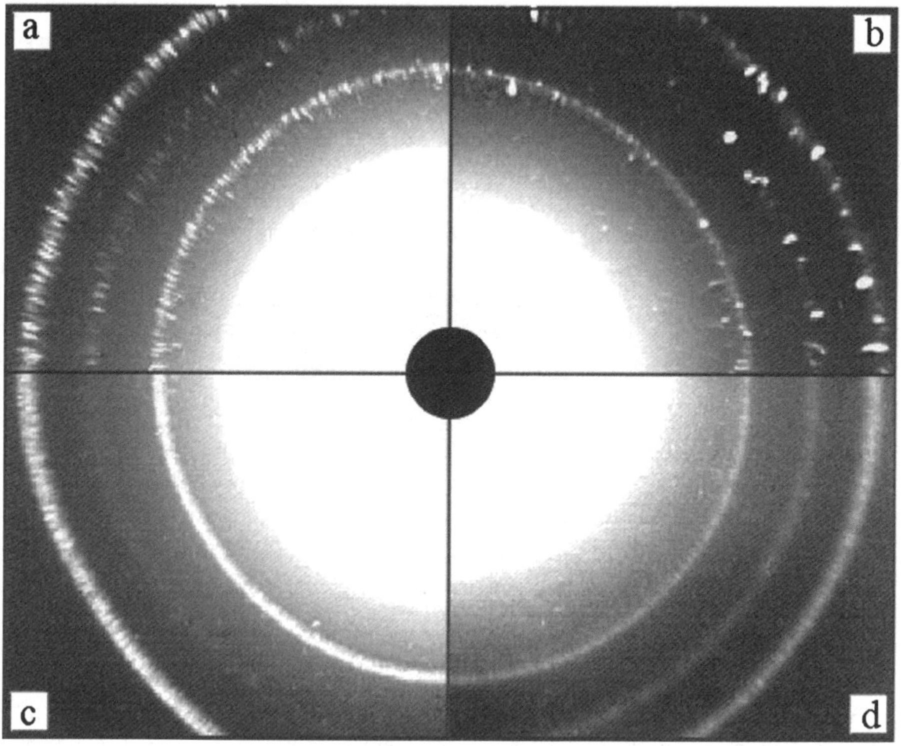

Figure 5.8 X-ray back reflection patterns of cobalt obtained from various powders hot pressed for 2 min at 950°C and 35 MPa: (a) Co Ultrafine (Umicore), 0.06 wt.% oxygen; (b) Co Extrafine (Umicore), in-mould reduced [12] to 0.16 wt.% oxygen; (c) Co Extrafine (Umicore), 0.44 wt.% oxygen; (d) Co 400 mesh (Umicore), 0.20 wt.% oxygen.

5.4 Phase Composition

The complexity of matrix compositions encountered in actual industrial situations, which are believed to aid either the manufacturing process or the tool performance [29], often leads to an absurd case when it is neither possible to control the system nor lay the foundations of its scientific understanding. It is simply not possible, within the limited length of this publication, to analyse the whole spectrum of matrix microstructures but it is indispensable to identify the key factors affecting the formation of various phases and the relations between the phase composition and properties of the matrix material.

The effects of powder characteristics and densification route on the microstructure and phase composition of cobalt have thoroughly been examined in several reviews [1,12,27,30–32]. Cobalt exists in two allotropic forms, a hexagonal close-packed (HCP) stable at temperatures below about 421°C, and a face-centred cubic (FCC) stable at higher temperatures [33]. The allotropic transformation has a martensitic nature and is associated with low free-energy change [33] which accounts for its sluggishness and sensitivity to the material processing conditions. In particular the grain size and degree of strain hardening achieved during consolidation influence the stability of the two allotropes, finer grain size and increased density of lattice defects favouring retention of the cubic form after cooling to room temperature.

The FCC structure (ε phase) transforms into the HCP one (α phase) by a slight shift in the atom locations which alters the stacking sequence of close-packed crystallographic planes [30]. Providing an external energy source, such as mechanical deformation, readily activates this process [30,32]. As exemplified in Fig. 5.9, shear stresses applied to the wear surface of a diamond-impregnated segment during stone sawing operations encourage the transformation.

The only means of preserving the FCC or mixed FCC/HCP structure in mechanically deformed cobalt is by alloying. For cast cobalt-base alloys the strongest cubic

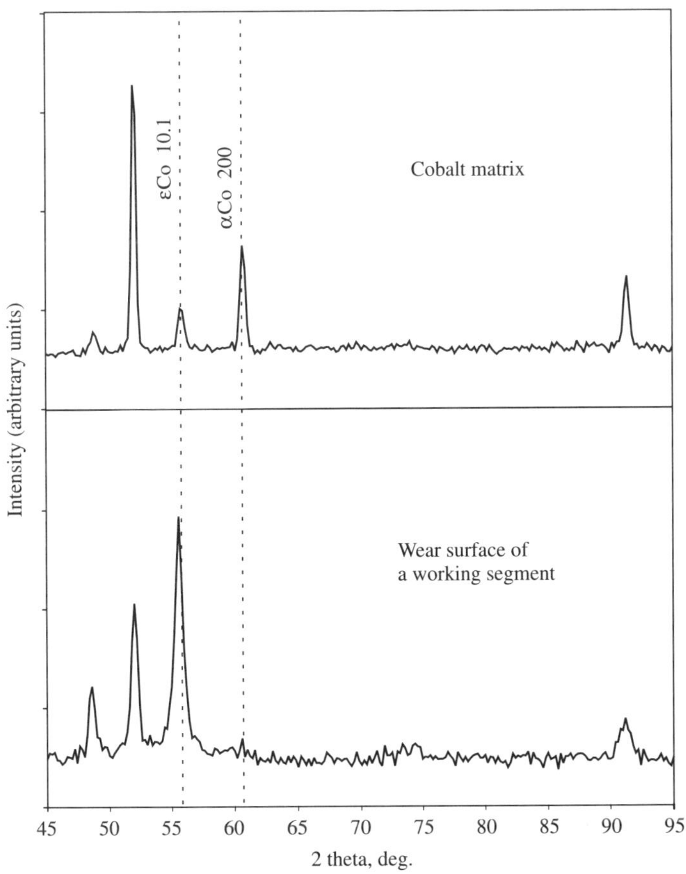

Figure 5.9 X-ray diffraction spectra recorded on as-consolidated specimen and working segment [32].

phase stabilisers have been identified and arranged in order of decreasing effect, viz. aluminium, niobium, titanium, iron, zirconium, tungsten and tantalum [34]. Other elements such as manganese, nickel, molybdenum, tin, and vanadium, are also reported to behave alike [35], although their ability to stabilise the α phase is weaker. In the PM materials aluminium, niobium, titanium and zirconium pose a hazard during consolidation since they combine easily with oxygen and therefore their response to stabilisation of the FCC structure is poor; whereas iron, especially in the form of a fine carbonyl powder, appears to be most suitable. In practice, addition of 10 wt.% iron to cobalt has been found extremely effective in stabilising the FCC structure, which generally results in marked improvement in the ductility of the material at the expense of its hardness and yield strength [12].

As it is seen from Fig. 5.10, alloys containing from about 10 to 24 wt.% iron have a thermodynamically stable two-phase structure, whereas further increase in the iron content yields a body-centred cubic (BCC) crystal structure which extends over the whole range of compositions up to pure iron.

It is important to note that the equiatomic cobalt–iron alloy undergoes a transformation from its disordered to its ordered state when cooled below 730°C. In practice,

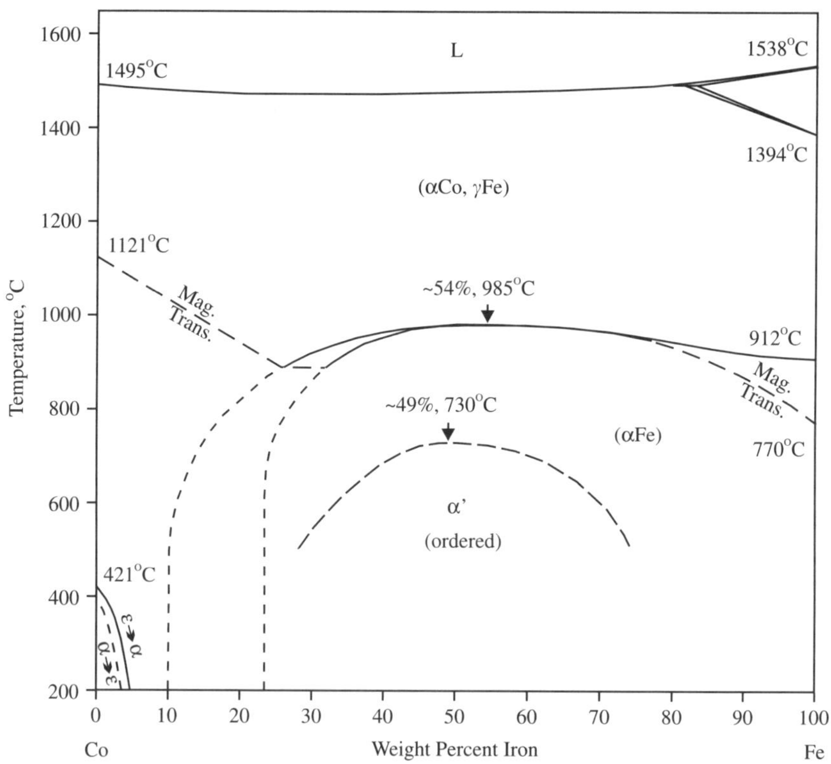

Figure 5.10 Phase diagram of the cobalt–iron system.

the ordered α' phase can exist over a considerable range on either side of the stoichiometric composition without becoming unstable.

Experimental observations of near-stoichiometric cobalt-iron compositions clearly indicate that the commercial powders are readily densified to virtually theoretical density by hot pressing at temperatures ranging from around 660°C for pre-alloyed powder [36] to 870°C [37] for pre-mixed material. However, ordering imparts brittleness to the material [37] thus imposing restrictions on its broader industrial application.

The fabrication of tool components for diverse non-critical applications involves the use of various bronze compositions. To date, tin bronzes combined with cobalt have gained the greatest practical significance. For a given matrix composition, the major distinction among the materials is the form of the starting bronze powder, i.e. a mixture of elemental powders of copper and tin, fully pre-alloyed powder, or a combination of these two. When pre-mixed elemental powders are used a transient liquid phase appears, when tin melts above 232°C, and interacts with the solid phases during the hot consolidation step leading to diffusional homogenisation of the alloy. If the material is consolidated by the hot pressing technique, the external pressure aids densification and pore collapse, but the short hold at relatively low temperature may be insufficient for creating a perfectly homogenised alloy. As demonstrated by the microstructures and X-ray diffraction spectra in Fig. 5.11, the microstructural inhomogeneity and marked peak broadening observed for the material made of mixed copper and tin powders indicate imperfect structure obtained in this case.

It is found in many systems that additives induce formation of intermetallic compounds in the consolidated material during its hot processing and cooling steps [25]. Intermetallic compounds usually nucleate at grain boundaries or interphase boundaries. As a consequence, they contribute to reduced mechanical properties. For this reason, in alloys of practical interest it is unfavourable if the minor components interact with the base metal to form brittle intermediate phases during the powder consolidation treatments.

Interestingly, the addition of pre-mixed bronze, containing 15 wt.% tin, to cobalt do not cause formation of detectable quantities of intermetallic compounds, although the hexagonal βCo_3Sn_2 phase is easily identified in a copper-free alloy containing 10 wt.% tin, as evident from Fig. 5.12.

The discussion up to now has mainly focused on the development of phase structure during hot pressing. In the case of tool components produced by the conventional cold pressing/sintering process the formation of various phases during the entire sintering cycle is critical because it governs the densification of the material.

In general, very fine powders are the preferred choice for the sintering process due to enhanced sintering properties of fine-grained and fine pore structures formed at the cold pressing step. The initial porosity is readily eliminated during sintering by vacancy diffusion from pores to the adjacent grain boundaries, which act as vacancy sinks, if the sintering parameters assure adequate diffusion rates and retention of fine-grained structure throughout the entire sintering period. This may well be exemplified by the consolidation of fine iron powders [38], which attain porosity of

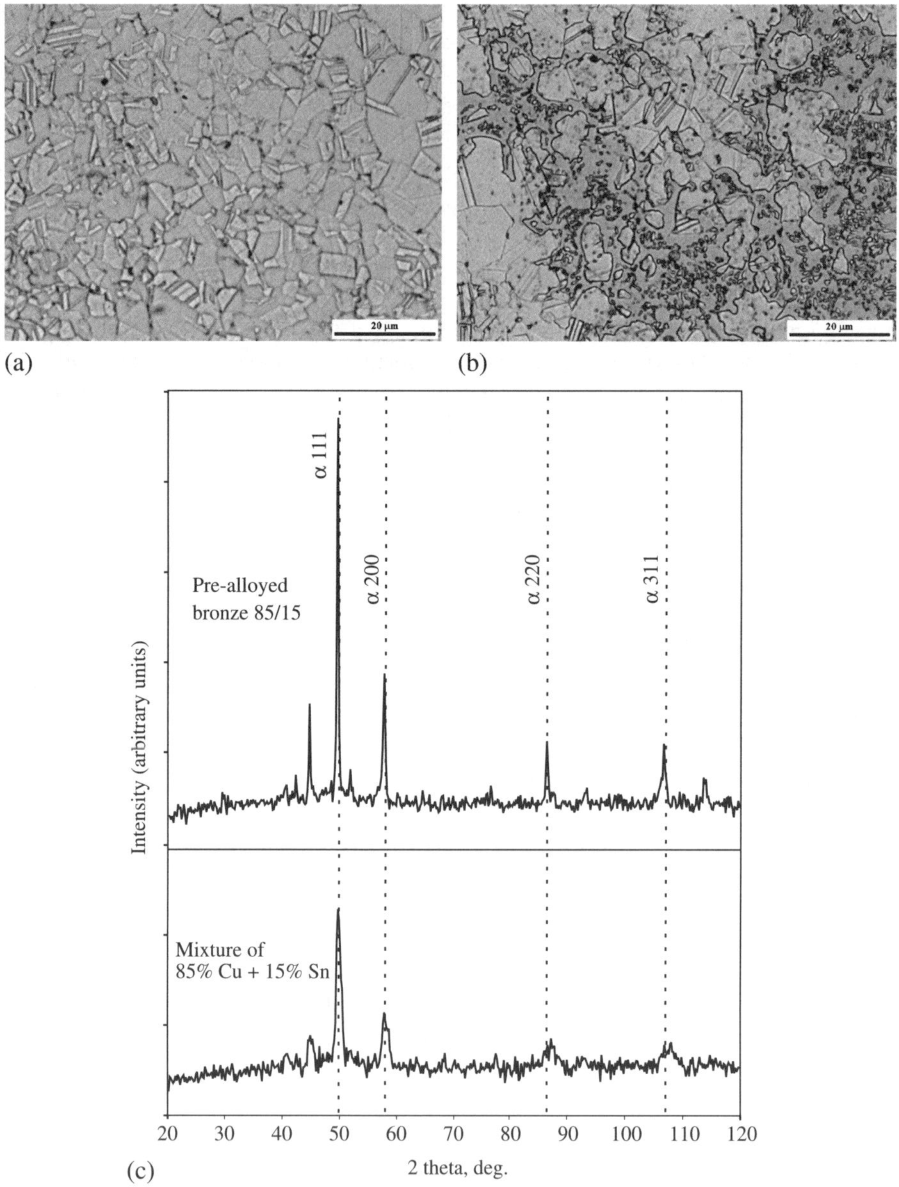

Figure 5.11 Microstructures of hot pressed tin bronze made of (a) pre-alloyed and (b) pre-mixed powders; (c) X-ray diffractograms.

around 4% after cold pressing at 490 MPa followed by sintering for 1 h at 900°C in hydrogen. An increase in the sintering temperature results in markedly lower final densities due to lower diffusion rates and a rapid grain growth in the γ phase (austenite) as compared with the α phase (ferrite). The grain growth isolates pores in the

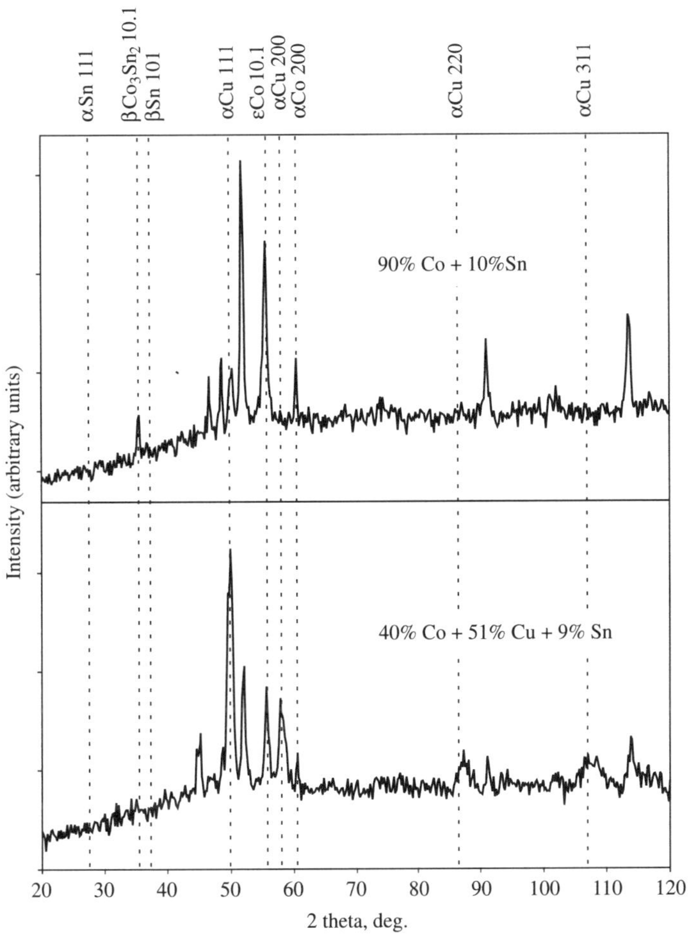

Figure 5.12 X-ray diffractograms for hot pressed cobalt–copper–tin and cobalt–tin alloys.

centre of large grains and slows down their elimination through the vacancy diffusion towards the grain boundaries.

The results obtained for fine pre-alloyed iron-copper powders corroborate the importance of the phase structure at the sintering temperature as the parameter controlling the rate of densification. Addition of copper to iron tends to stabilise the austenite, which has the FCC structure, and therefore higher shrinkage rates are observed within the ferrite field, i.e. in the BCC α phase, than above the eutectoid temperature of 850°C [21], as is exemplified by the sintering curve for Next 900 in Fig. 5.4.

If alloying is necessary, pre-alloyed powders are generally better suited for sintering than their pre-mixed counterparts. In the latter case, diffusional homogenisation of the sintered material often leads to pore formation by the Kirkendall effect due to unequal diffusion rates of the alloy-forming elements [39]. Additionally, if any of the mixed constituents melts to form a transient liquid phase on heating to the sintering temperature, large pores may result from solubility of the liquid in the solids thereby contributing to reduced mechanical properties of the sintered material. This has been observed in sintered tin bronzes obtained from pre-mixed powders where the tin melts and spreads into the copper structure thus generating pores at the prior tin particle sites [39,40].

References

1. Konstanty, J., Bunsch, A., Hot pressing of cobalt powders. Powder Metallurgy, 1991, 34(3), 195–198.
2. Cobalt. Technical information for diamond tool producers. Union Miniere, May 26, 1993.
3. Höhne, M., Co in diamond tools. Paper presented at Seminar on PM Diamond Tools, Lausanne, Switzerland, November 2–3, 1995.
4. Hot press properties. Fine cobalt powders. Sherritt Inc., Technical information, December 1995.
5. Hot press properties. Fine cobalt powders. The Westaim Corporation, Technical information, October 1996.
6. Cobalt powders for diamond tools. Union Miniere, Technical information brochure.
7. Metal matrix powders for diamond tools. Union Miniere, Diamond Tool Brochure – DT/All Products/E/0998.
8. UM cobalt and energy products. Diamond tool product line. Union Miniere, Technical information, January 5, 2000.
9. Rigby, P., Peersman, J., Kamphuis, B.J., Huh, S., Mishra, P., Recent developments in bond powders for diamond tools. Union Miniere internal paper.
10. UM cobalt and energy products. Diamond tool market. Union Miniere, Technical information.
11. Sub Micron Sized (SMS) cobalt powder. Sintering data for granulated and non-granulated product. Union Miniere technical information.
12. Konstanty, J., Cobalt as a Matrix in Diamond Impregnated Tools for Stone Sawing Applications. AGH Uczelniane Wydawnictwa Naukowo-Dydaktyczne, 2nd Edition, Krakow, 2003.
13. Next 100. Eurotungstene Metal Powders. Product information brochure, October 1, 1996.
14. Next 200. Eurotungstene Metal Powders. Product information brochure, October 1, 1996.
15. Next 300. Eurotungstene Metal Powders. Technical data sheet, Version 1, May 2003.
16. Next 900. Eurotungstene Metal Powders. Technical data sheet, January 2004.
17. Clark, I.E., Kamphuis, B.J., Cobalite HDR – a new prealloyed matrix powder for diamond construction tools. Industrial Diamond Review, 2002, 62(3), 177–182.
18. Kamphuis, B.J., Serneels, A., Cobalt and nickel free bond powder for diamond tools: Cobalite CNF. Industrial Diamond Review, 2004, 64(1), 26–32.
19. del Villar, M., Muro, P., Sánchez, J.M., Iturriza, I., Castro, F., Consolidation of diamond tools using Cu–Co–Fe based alloys as metallic binders. Powder Metallurgy, 2001, 44(1), 82–90.
20. Cai, O., Sintering without pressure, or free sintering. Diamante Applicazioni & Tecnologia, 1997, 3(11), 106–122.
21. Binary Alloy Phase Diagrams. Edited by Massalski, T.B., 2nd Edition, Vol. 2, ASM International, 1990, pp. 1232–1233, 1408–1410.
22. Type A ultrafine cobalt powder. The Westaim Corporation. Technical information, TB010.DOC.
23. German, R.M., Sintering Theory and Practice. Wiley, New York, 1996.
24. Gao, J., Thompson, R.G., Patterson, B.R., Computer simulation of grain growth with second phase particle pinning. Acta Materialia, 1997, 45(9), 3653–3658.
25. German, R.M., The use of phase diagrams in predicting sintering behavior. In Horizons of Powder Metallurgy, Part II, edited by Kaysser, W.A. and Huppmann, W.J., Verlag Schmid GmbH, Freiburg/Germany, 1986, pp. 1239–1242.

26. Blicharski, M., Gorczyca, S., Rekrystalizacja z udziałem drugiej fazy. Wydawnictwo Śląsk, Katowice, 1980.
27. Buekenhout, L., Berghezan, A., A new approach to the problem of room temperature ductility of polycrystalline cobalt. In Proceedings of 10th Plansee Seminar, Reutte, Austria, 1–5 June, 1981, Vol. 2, pp. 899–916.
28. Robino, C.V., Representation of mixed reactive gases on free energy (Ellingham–Richardson) diagrams. Metallurgical and Materials Transactions B, 1996, 27B, 65–69.
29. Chalkley, J.R., Thomas, D.M., The tribological aspects of metal-bonded diamond grinding wheels. Powder Metallurgy, 1969, 12(24), 582–597.
30. Berghezan, A., Beukenhout, L., A critical review of the ductility of polycrystalline cobalt and its experimental approach via powder metallurgy. In Proceedings of the International Conference on Cobalt, Brussels, Belgium, November 10–13 , 1981, Vol. 2, pp. 295–307.
31. Akyüz, D.A., Doctoral Thesis, EPFL, Lausanne, 1999.
32. Romański, A., Doctoral Thesis, AGH, Kraków, 2000.
33. Betteridge, W., Cobalt and its Alloys. Ellis Horwood, Chichester, 1982.
34. Diderrich, E., Drapier, J.M., Coutsouradis, D., Habraken L., Low-alloy ductile cobalt for hardfacing electrodes. Cobalt, 1975, (1), 7–16.
35. Zhao, J.-C., The fcc/hcp phase equilibria and phase transformation in cobalt-based binary systems. Zeitschrift für Metallkunde, 1999, 90(3), 223–232.
36. Mende, B., Gille, G., Gries, B., Aulich, P., Münchow, J., Pre-alloyed powder. U.S. Patent 6,554,885 B1 (April 29, 2003).
37. Konstanty, J., Ratuszek, W., Jamrozek, J., Olszewska, I., Properties of hot pressed near- stoichiometric FeCo alloys. International Journal of Powder Metallurgy, 2004, 40(1), 41–51.
38. Hickling, H., Coleman, D.S., Sintering of fine iron powders produced from ferrous oxalate dihydrate. Powder Metallurgy, 1982, 25(1), 25–34.
39. German, R.M., Liquid Phase Sintering. Plenum Press, New York, 1985.
40. Peissker, E., Metal Powders. Norddeutsche Affinerie, Hamburg, June 1986.

CHAPTER 6

Mechanical Properties of the Matrix

An important requirement of the matrix is to firmly hold the diamond particles. Under the application conditions the diamond–matrix interactions occur in a variety of forms depending on the size and shape of each individual diamond particle, its orientation and loading conditions, residual stresses in the matrix, diamond–matrix friction, etc. The existing theoretical knowledge of diamond retention has evolved from simplistic models and can hardly give a satisfactory explanation of the complexity of individual diamond pullout events observed in the industrial practice.

The actual mechanical response of the matrix subjected to an application of a pulsing force of varying intensity is apparently affected by its fatigue properties under complex loading conditions and at moderately elevated temperature. Further complications arise from internal stresses set up around each diamond particle during the manufacturing process due to mismatched thermal expansion coefficients. These stresses are believed to enhance retention, but direct quantification of their contribution presents difficulty.

It is important that this pre-stressed state is not annihilated by plastic deformation or brittle failure of the matrix, caused by external forces applied through the working diamonds, or due to reverse thermal expansion of the matrix caused by the heat generated in the cutting zone. Otherwise the diamonds may become loose, as demonstrated in Fig. 6.1, and fall off prematurely.

Certain material characteristics such as hardness, yield strength and impact strength, are generally believed to control the matrix capacity for diamond retention. Although no consensus has been reached to date on practical relevance of the above-mentioned retention indicators to various application conditions, evaluation

Figure 6.1 Evident debonding between the working diamond and matrix.

of hardness, yield strength, bending strength and impact strength has become a routine industrial practice.

6.1 Hardness

A properly densified matrix acquires a narrow hardness range which, to a great extent, is affected by its composition and parameters of the powder consolidation process. In the case of incomplete densification or faulty processing of the matrix–diamond mixture the hardness readings do not fall within the specified range. Thus, hardness mainly becomes a usable quality control parameter.

A typical relationship between hardness of the matrix and its consolidation temperature is demonstrated by curve 1 in Fig. 6.2.

At lower consolidation temperatures hardness increases in direct proportion to the amount of plastic deformation and density of the material. When full density is approached then other processes like recrystallisation and grain growth can be activated to soften the material. From the quality control standpoint it is desirable for the tool fabrication that the matrix powder exhibits an elongated plateau on the hardness vs. consolidation temperature plot, as exemplified by curve 2 in Fig. 6.2.

The actual response of hardness to hot pressing temperature for various commercial cobalt powders is presented in Fig. 6.3.

In the tool-making practice, hardness has also been used as a convenient guide to the diamond-holding power of fully consolidated matrices. This simple criterion ignores, however, a few important facts. First, hardness measurements involve plastic strain of between 5% and 12% and therefore the hardness readings, being affected by the strain hardening characteristics of the tested material, cannot be

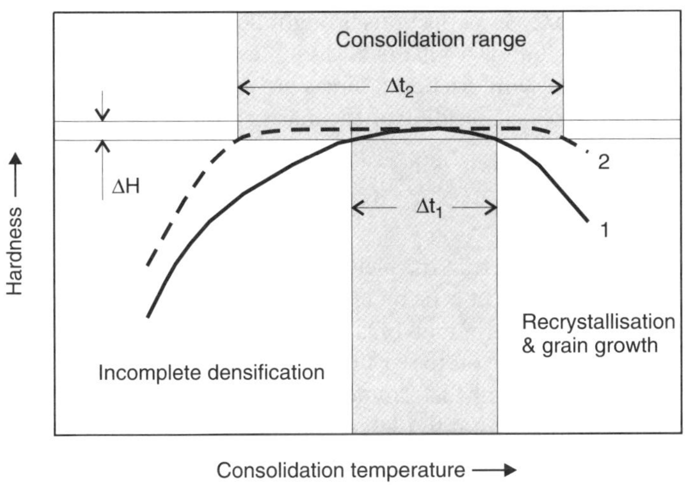

Figure 6.2 Hardness as a function of powder consolidation temperature.

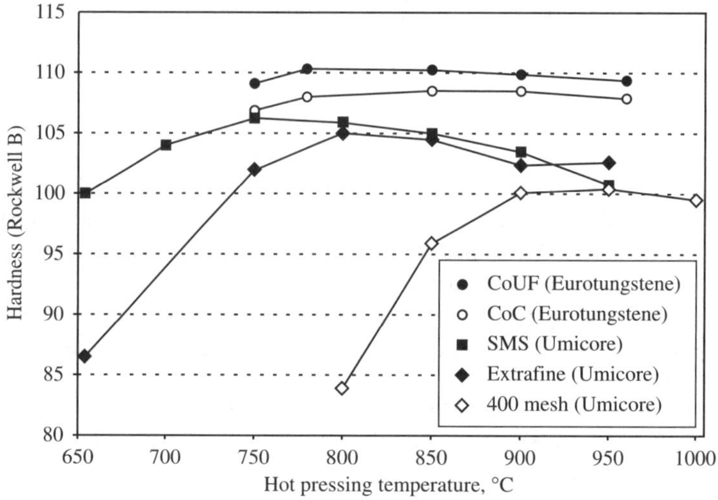

Figure 6.3 Dependence of hardness on hot pressing temperature for selected commercial cobalt powders [1–3].

directly correlated with its elastic behaviour. Second, in certain applications, e.g. in circular sawing, the diamond/matrix set-up is subjected to a high-frequency impact loading therefore the matrix sensitivity to the loading conditions becomes another important parameter, which cannot be deduced from any static test. And finally, when the matrix deforms it slips along the face of the diamond (see Section 3.3.1). The static coefficient of friction for diamond on most metals is small and therefore the metallic matrix is nearly free to slip laterally at the interface. The situation changes when the coefficient of friction is markedly increased due to coatings on diamond or diamond surface imperfections brought about either by the diamond manufacturing conditions or by a chemical attack exerted by the matrix during its consolidation. Thus growing frictional forces are expected to increase the pressure necessary to produce plastic yielding of the matrix around the working diamond.

6.2 Yield Strength

It has been postulated [4–6] that the yield strength is another important property of the matrix, which has a direct bearing on its capacity to retain diamond grits. Each time the yield strength of the matrix is exceeded, the seat of the diamond opens slightly and thereby the hold on the grit is gradually destroyed.

The yield strength is usually determined in the tensile test and, as with hardness, the results may prove irrelevant to high strain rate conditions which are typical for certain applications, e.g. circular sawing. Except for certain steels, which display a double yield point, the stress at which the material starts to deform plastically is not easily detected. Therefore, an offset yield strength, describing the stress that gives typically 0.2% of plastic deformation, is usually measured.

As it has been thoroughly discussed in Sections 5.2 and 5.3, the amount and dispersion of the oxide phase in conjunction with the powder particle size and its consolidation conditions govern the occurrence of recrystallisation and the final grain size of the matrix. Consequently, the tensile characteristics of the material may flexibly be engineered by altering the powder properties and/or its processing route as exemplified with cobalt in Fig. 6.4. The powders have been consolidated at a relatively high temperature so as to contrast their susceptibility to grain growth and to demonstrate the influence of impurities, such as oxygen and sulphur (see Section 3.2.1), on tensile properties.

The highest yield strength is obviously obtained in a fine-grained material containing high amount of finely dispersed, stable oxides (Curve I). On the other hand, the yield strength of cobalt is drastically reduced when either the powder is consolidated by sintering in a reducing atmosphere (Curve III) or a non-oxidised powder is subjected to hot pressing (Curve IV).

The percentage elongation generally increases as the yield strength decreases except in situations where other factors, e.g. a presence of sulphur, reduce ductility of the material.

It is worthwhile to mention that further diversification of the yield strength is possible by admixing other powders to the base metal prior to consolidation [7].

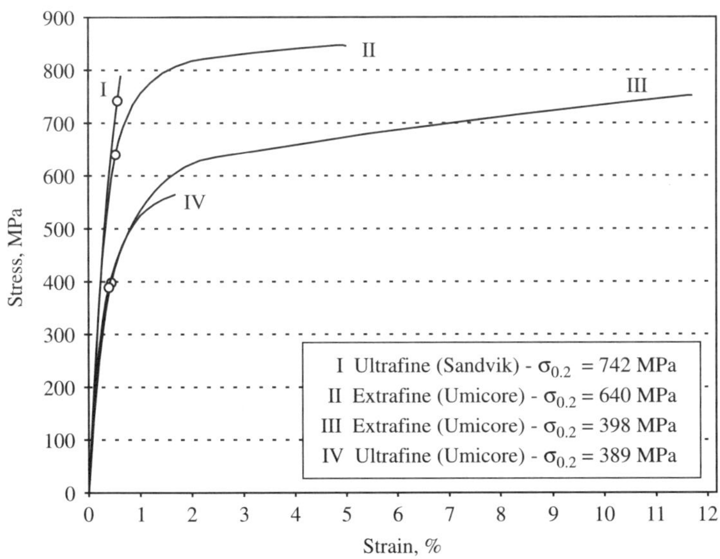

Figure 6.4 The stress–strain curves for cobalt specimens hot pressed for 2 min at 950°C and 35 MPa (Curves I–III), compared to a specimen sintered for 1 h at 950°C in hydrogen atmosphere (Curve IV).

Specifically fine tungsten carbide is commonly used to increase the yield strength by converting the matrix to a particulate-reinforced composite.

6.3 Bending Strength

Irrespective of methodological incorrectness and problematic interpretation of results acquired outside the elastic range, the bending test has become a widespread screening procedure whereby the material's as-sintered integrity is assessed in a simple manner.

In the industrial practice, the bending test has also proved useful for assessing the effectiveness of various coatings deposited on diamond in enhancing its retention in the matrix [8]. In comparison with specimens containing uncoated diamond, the bending strength of specimens impregnated with coated diamond grits, of the same size and concentration, should reach higher level and, more importantly, the percentage of broken coated crystals across the break of the test specimen should be markedly higher compared to uncoated ones, thus providing the toolmaker with a quantitative indication that the tool composition and consolidation conditions have been correctly chosen.

6.4 Impact Strength

As mentioned in the previous sections, the working diamonds periodically experiencc intense impacts that induce stresses in the surrounding matrix. Besides, diamond tools are generally liable to abuse and therefore undesirable vibrations will further add to the impact loading of the matrix degrading its ability to hold the diamonds firmly. Such incidents are encountered with diamond particles embedded in a too brittle matrix that starts to break away rather than wear away. Therefore, in the vast majority of applications, the impact strength of the matrix becomes critical to the tool performance.

The impact strength of PM materials is influenced in a complex manner by their density, chemical and phase composition, grain size, contents of impurities, etc. It is essential that the material is fully densified, otherwise its toughness declines dramatically with increasing porosity [7,9].

The energy required to fracture unalloyed cobalt specimens, obtained by hot pressing of powders finer that 2 μm (Fisher sub-sieve size) to virtually full density, is mostly affected by impurities contained in the material. As demonstrated in Fig. 6.5, the ductility of cobalt is apparently controlled by the degree of oxidation and contamination with sulphur.

The specimens made of sulphur contaminated hydrometallurgical powders (see Section 3.2.1) yield very low impact strength figures despite the absence of oxygen. In the near sulphur-free material the impact strength sharply declines as the oxygen content is increased from below 0.01 to about 0.2 wt.%. Beyond this range, the impact strength slopes down more gently. The increase in hot pressing temperature from

800°C to 950°C has a marginal effect and seems to be restricted to very low oxygen contents (~0.01 wt.%), where the rise in temperature stimulates ductile behaviour.

The impact response records obtained from an instrumented test reveal the effect of oxygen on the failure history of the material. As shown in Fig. 6.6, the specimen

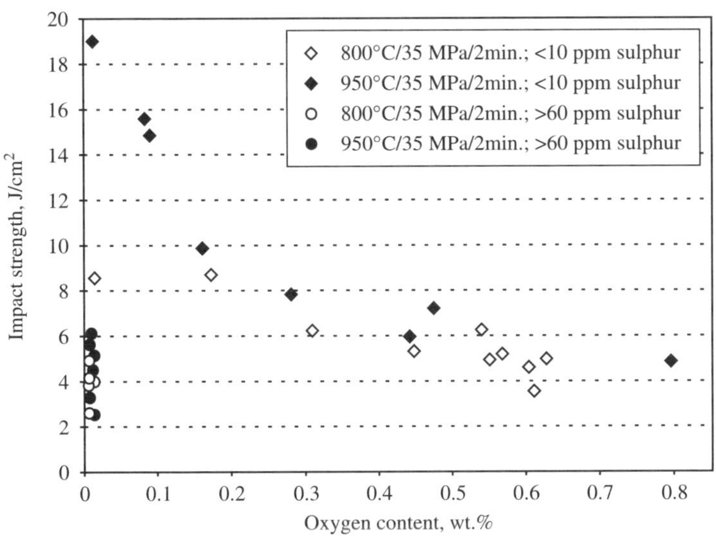

Figure 6.5 Combined effects of oxygen content and contamination with sulphur on impact strength of hot pressed cobalt.

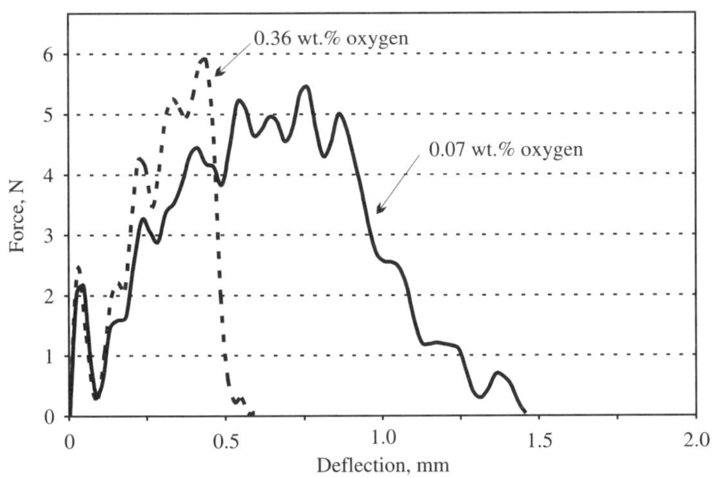

Figure 6.6 The force-deflection impact response traces of cobalt specimens containing various amounts of oxygen [10].

containing 0.36 wt.% oxygen breaks brittle in the macroscopic elastic field, but ranks high in ability to absorb force. On the other hand, the low-oxygen cobalt (0.07 wt.%) yields at lower force but undergoes sizeable plastic deformation consuming markedly more energy for plastic deformation and crack propagation.

From the diamond retention standpoint, it seems desirable that the material has the ability to absorb some deflection, otherwise cracks may be initiated and propagated in the matrix as the result of large strains generated in the vicinity of sharp diamond edges [11].

References

1. Technical data sheet CO7106. Eurotungstene metal powders, March 2004.
2. Technical data sheet CO6101. Eurotungstene metal powders, March 2004.
3. Metal matrix powders for diamond tools. Union Miniere, Diamond Tool Brochure – DT/All Products/E/0998.
4. de Châlus, P.A., Metal powders for optimum grain retention. Industrial Diamond Review, 1994, 54(4), 170–172.
5. Chalkley, J.R., Thomas, D.M., The tribological aspects of metal-bonded diamond grinding wheels. Powder Metallurgy, 1969, 12(24), 582–597.
6. Bonneau, M., Tensile rupture on sintered cobalt specimens. In Proceedings of Seminar on PM Diamond Tools, Lausanne, Switzerland, November 2–3, 1995, pp. 41–47.
7. Konstanty, J., Cobalt as a Matrix in Diamond Impregnated Tools for Stone Sawing Applications. 2nd Edition, AGH Uczelniane Wydawnictwa Naukowo-Dydaktyczne, Krakow, 2003.
8. McEachron, R., Connors, E.J., Slutz, D.E., Multi-layer metal coated diamond abrasives with an electrolessly deposited metal layer. U.S. Patent 5,232,469 (August 3, 1993)
9. Konstanty, J., The materials science of stone sawing. Industrial Diamond Review, 1991, 51(1), 27–31.
10. Lenkey, G.B., The instrumented impact experiments on Co-based materials. University of Miskolc, 20 February 1998 (unpublished).
11. Dudek, K., Pietrzyk, M., Konstanty, J., Analiza stanu naprężenia w otoczeniu kryształu diamentu w segmencie metaliczno-diamentowym. In Proceedings of 10th Conference on Computer Science in Metal Technology, Wisła Jawornik, Poland, 12–15 January 2003, pp. 95–102.

CHAPTER 7

Wear Properties of the Matrix

A number of fundamental studies of diamond tool wear processes may be found in the literature [1–18]. The majority of them have been entirely devoted to the behaviour of diamond grits, whereas little attention is turned to the mechanisms involved in the wear of the matrix. Therefore, it is worthwhile to discuss the complexity of the matrix wear, which seems to be equally a system property and material characteristics.

During sawing, or grinding, with diamond tools the interactions between the workpiece debris and matrix occur in a variety of forms depending on size of the abrasive swarf, its shape, cleavage properties, hardness, loading conditions, particle movement speed and pattern, etc. Abrasive and erosive modes of wear are always accompanied by other mechanisms such as thermal fatigue, indentation fatigue and others which are not yet well understood.

7.1 Resistance to Abrasive Wear

As stated above, there are always several mechanisms of wear acting in concert, all of which have different characteristics. The abrasive particles may remove material by micro-cutting, micro-fracture, pullout of individual grains or fatigue by repeated deformations [19]. Considerable plastic deformation occurs beneath the abraded surface and, consequently, strain-hardening takes place in the material which usually results in a reduction of the abrasive wear rate. As shown in Fig. 7.1, severe subsurface deformation is inevitable during abrasion and therefore the hardness at high strains often becomes the controlling property [18–20].

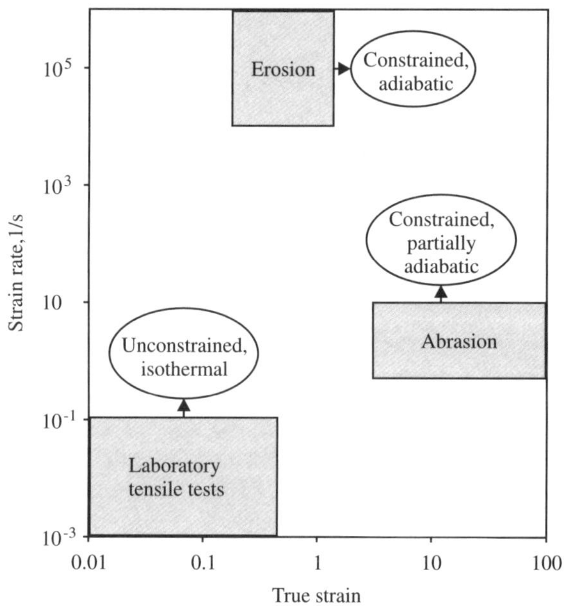

Figure 7.1 The strain rate–strain regime associated with erosion, abrasion and a quasi-static tensile test [20].

If hardness of the work-hardened material is close to or exceeds that of the abrasive, the reinforcement produced by strain-hardening is effective and the wear resistance is increased [19,21,22]. From the viewpoint of effective cutting action, the abrasive must be harder than the wearing surface. If it is only a little harder than the material to be abraded, the abrasive tends more readily to have its sharp corners removed after which its own cutting action is drastically reduced.

The mechanism of wear is affected to a large extent by the shape of abrasive particles. Abrasives originating from freshly fractured material have a lot of micro-cutting edges and remove far more material than the worn ones, which have only rounded edges. If the abrasive material is too brittle, however, then it may break into fine particles thus minimising wear. At constant contact pressure, and other testing conditions, the wear rate of most metals increases non-linearly with the mean abrasive particle size up to about 50 μm and reaches a limiting value at about 100 μm [19,23–25].

Another factor controlling wear rates is the tendency for material to be displaced rather than removed as wear debris. Therefore, outstanding ductility and toughness can suppress the more rapid forms of abrasive wear, such as micro-cutting and brittle fracture. Indentation fatigue, which is a much slower wear process, can appear instead. Needless to mention that the sharpness of abrasive particles and the pattern of moving, i.e. sliding or rolling, will also have an important effect on the process of material displacing.

Identification of minerals in the workpiece, that cause abrasion, is also an important consideration. Minerals that are too soft to abrade may still wear the material but the mechanisms involved are different, e.g. thermal fatigue, oxidation and removal of oxide layers by abrasion, etc.

Abrasive wear tests conducted on a great number of metallic and non-metallic materials have indicated that when the ratio of the hardness of the abrasive to the hardness of the test material attains the value of 1.3–1.7, the relative wear has a definite maximum and constant magnitude [21]. The foregoing finding implies a possibility of suppressing abrasion by raising hardness of the material above a certain critical value. The simplest way of accomplishing this is by introducing hard phases, such as carbides, to a softer matrix. A material containing carbides can possess up to four times the abrasive wear resistance of the corresponding carbide-free single-phase material. The peak of abrasive resistance occurs at approximately 30 vol% of hard phase. Beyond this level an undesirable, brittle and continuous hard phase network appears to facilitate crack propagation throughout the structure and to reduce resistance to wear [19,26]. If the abraded material also experiences impact stresses, its ductility takes priority over any other properties.

The way the grits pass over the worn surface determines the nature of abrasive wear. The literature denotes two basic modes of abrasive wear, i.e. two- and three-body abrasive wear.

7.1.1 Two-Body Abrasive Wear

The two-body abrasive wear is well exemplified by the action of grinding paper on a surface. Fixed abrasive grits pass over the mating surface and cut groves in it.

The two-body abrasive wear is the most rapid and severe means of removing material and therefore it is of the utmost importance that the diamond protrusion ensures sufficient clearance to keep the matrix remote from the workpiece surface asperities. As exemplified in Fig. 7.2, it seems impossible to avoid incidental situations when the matrix locally comes into direct contact with the workpiece and is abraded by the two-body mechanism.

If the clearance is insufficient for the application, repeated exposures of the matrix to a direct contact with the processed material will inevitably lead to a drastic decrease in the useful life of the tool.

7.1.2 Three-Body Abrasive Wear

The three-body abrasive wear arises when hard abrasive particles are introduced between a pair of sliding surfaces and abrade material of each. The particles are free to roll as well as slide over the surfaces since they are not held rigidly. It was found that the three-body abrasive wear is approximately 10 times slower than its two-body counterpart because the abrasive particles spend about 90% of the time rolling and only about 10% of the time sliding and abrading the surfaces [22,23].

Typically, the three-body mode of wear does not occur when the abrasive particles are small, or when they are softer than the abraded material. The latter may not

Figure 7.2 A series of parallel scratches produced in the matrix by stone asperities sliding over the side face of a sawblade segment.

apply to coarse abrasive particles, which neither are rounded up nor flow plastically but tend to shatter under increased loading thus reforming very sharp cutting facets. Each particle destruction event is accompanied by formation of new surface, which requires expenditure of energy [27]. Since this energy is provided via the wearing material, its surface is inevitably damaged while splitting the abrasive particles [26].

In sawing stone, concrete and hard ceramics, for instance, the wear of the matrix is brought about by the passage of an abrasive slurry, being a mixture of workpiece debris and a carrying liquid, over the working face of the diamond-impregnated segment. The abrasivity of the swarf, internally generated in the cutting zone by working diamonds, is influenced in a complex manner by its properties, segment composition and sawing conditions. The most important workpiece debris characteristics are:

- average particle size and size distribution;
- particle shape (sharpness);
- mineral composition (hardness);
- cleavage properties (resistance to fragmentation).

The variables related to segment composition and sawing conditions are equally important as they affect the space by which the matrix clears the workpiece and control the swarf concentration in the slurry. Hence the other important considerations are:

- diamond size, strength and concentration (which combine to affect its protrusion window);
- maximum chip thickness (see Eqs. (2.1) and (2.3));
- cumulative chip thickness (see Section 2.1).

All the above parameters govern the particle movement pattern. Figure 7.3 shows a simplified two-dimensional model of a tribological system wherein some portions

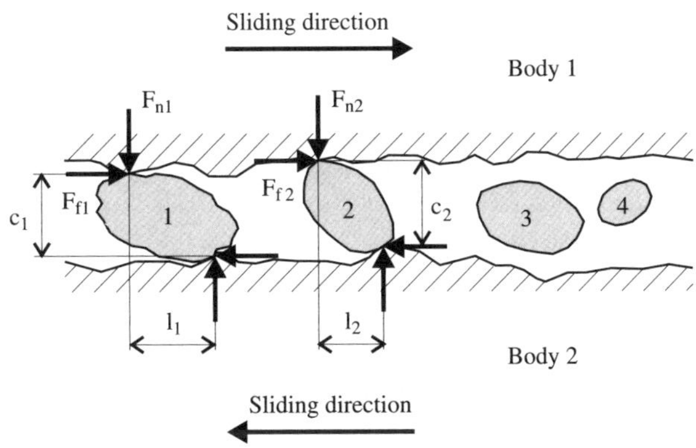

Figure 7.3 Schematic representation of forces acting on abrasive particles.

of an intermediate matter are compelled to either sliding or rolling motion between two sliding bodies.

Ignoring fragmentation of abrasive particles, it is assumed that torques produced by the frictional forces F_f tend to make the particles roll, whereas counteracting torques brought about by the normal forces F_n tend to make the particles slide. Hence rolling occurs when

$$F_f c > F_n l \tag{7.1}$$

whereas sliding occurs when

$$F_f c \leq F_n l \tag{7.2}$$

It is convenient to replace F_f/F_n by a coefficient of friction μ. Equations (7.1) and (7.2) will then become

$$\mu_r > l/c \tag{7.3}$$

$$\mu_s \leq l/c \tag{7.4}$$

where μ_r and μ_s are coefficients of friction for the rolling and sliding particle, respectively.

From Eqs. (7.3) and (7.4) it can be inferred that the ratio of the spacing l between the indentation forces to the spacing c (clearance) between the frictional forces becomes an important tribological parameter which determines the abrasive particle movement. The ratio l/c depends on the particle size and shape, normal load, hardness and hardness ratio of the two mating surfaces [28].

Obviously, a large particle size and small clearance favour sliding movement (see particle 1 in Fig. 7.3) thereby drastically increasing the rate of wear. This has been confirmed experimentally by sawing sandstone with a circular sawblade that was previously conditioned on granite [18]. The markedly coarser stone debris combined with a threefold increase in the cutting rate, leading to deeper workpiece penetration by diamond particles as described by Eq. (2.1), has initiated a severe matrix abrasion by the sliding action of coarse particles. Figure 7.4 shows the evolution of diamond protrusion and evidence of harsh abrasion on the working face of the sawblade.

As the particle size to clearance ratio is sufficiently lowered to attain a certain critical value, the particle rolls (see particle 2 in Fig. 7.3). It is also possible that some particles momentarily lose contact with either surface (particle 3) or are too small to abrade (particle 4).

Modelling of the matrix wear rates has long been recognised as one of the most important tool-making issues. Laboratory wear testing is without doubt the quicker and relatively inexpensive means of obtaining information on wear rates and wear mechanisms. It must be realised, however, that laboratory experiments, even when conducted with highly sophisticated testing rigs, inevitably limit the number of variables and incompletely imitate the complex processes involved in the actual tool application environment. Another complication is the proper selection of the backing

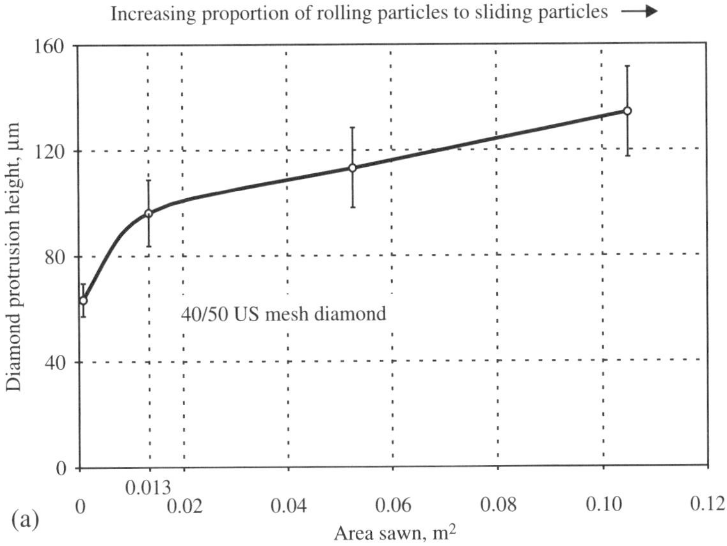

(a)

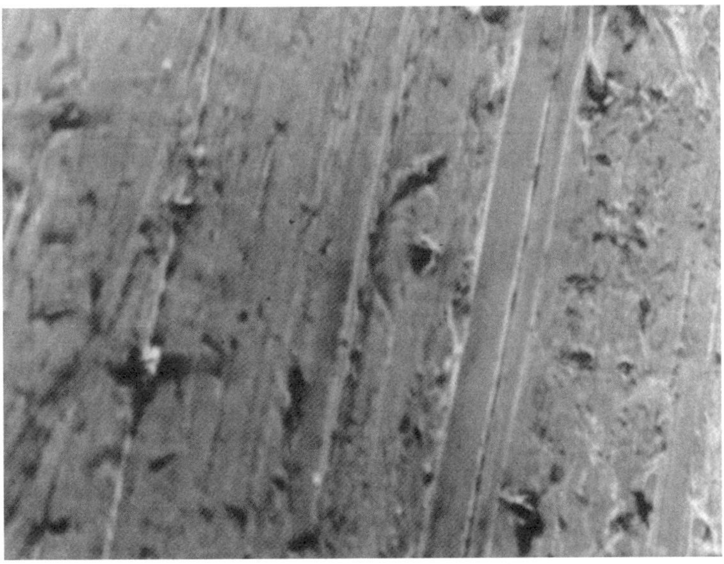

(b)

Figure 7.4 Evolution of diamond protrusion (a) and the worn surface of the tool (b) after sawing 0.013 m² of sandstone.

wheel that forces the abrasive particles into the abraded material. There is sufficient experimental evidence that hardness and surface topography of the backing wheel appear to be extremely important for three-body abrasive wear [18,19]. As demonstrated in Fig. 7.5, the wear rates evaluated on different backing wheels do not remain in proportion to one another and materials are ranked differently.

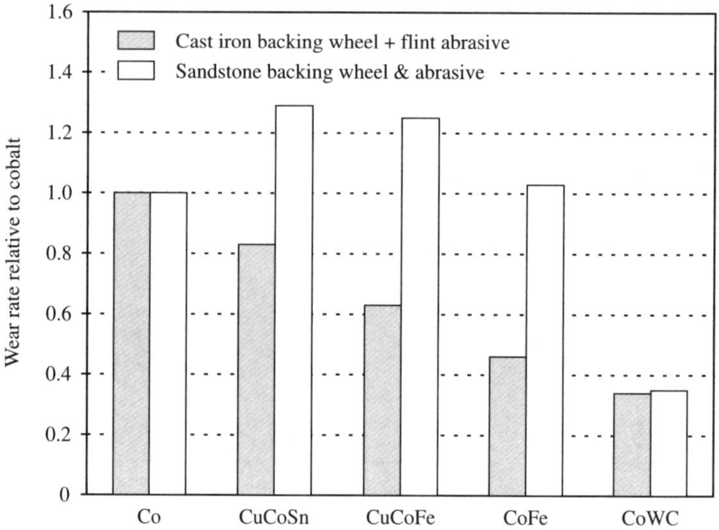

Figure 7.5 The effect of backing wheel on wear rate of various matrix materials [18].

Undeniably, wear resistance of the matrix is one of the most important properties in assessing the tool serviceability in any specific application. Although diverse test equipment is in current use [18,23,29–40] no simple technique has yet been developed to simulate the tool application conditions which are characterised by a large number of wear variables, such as load, type of motion, velocity, contact area and geometry, sliding distance, temperature and cooling conditions, vibration, workpiece (debris) characteristics, etc. Therefore, additional refinements to the existing laboratory wear testing procedures are required so as to better reproduce the tool wear conditions.

7.2 Resistance to Erosive Wear

Erosive wear is caused by the impact of solid particles against the surface of an object. It involves several mechanisms, which are largely controlled by the particle material, particle size, angle of impingement, impact velocity and erosion medium.

In a similar manner to abrasive wear, hard particles cause higher wear rates than soft particles. It is, however, impossible to isolate hardness completely from other features of the impacting particle such as its shape and resistance to fragmentation. Even if the particle is hard but relatively blunt or spherical then it is unlikely to cause severe erosive wear. On the other hand, when the particle is sharp then cutting or brittle fragmentation of the target material is more likely.

The size of the particle is also of considerable relevance. The efficiency of erosive wear sharply decreases as the particle size decreases below about 100 μm [41]. For instance, quartz particles smaller than 5 μm in size are reported to cause no measurable erosion of steel at impact velocities below 300 m/s [42].

The angle of particle impingement can range from 0° to 90°. At small angles, up to around 4°, the rate of erosion is negligibly low or even undetectable [43]. In ductile metals, measurable wear occurs as the angle of incidence increases and reaches its maximum rates at angles between 15° and 30° [19,20,41,44]. Schematic representation of the effect of impingement angle on wear rate of ductile materials is shown in Fig. 7.6.

A theoretical analysis predicts that beyond the angle for maximum erosion the wear rate, represented by the solid curve in Fig. 7.6, slopes down to reach zero at 90° [44]. The idealised model considers, however, a single particle collision with a perfectly flat surface, whereas in actual situations the surface becomes roughened by many impacts. Therefore, subsequent particles strike the surface with a distribution of angles deflecting upwards the experimental curve represented by the dashed line in Fig. 7.6. A number of other mechanisms such as brittle behaviour due to work-hardening, fragmentation of the impacting particles, low cycle fatigue, temperature effects due to high strain rates, delamination wear, etc. have also been suggested for material removal due to impingement at high angles [41].

Another factor, which has a very strong effect on the wear process is the impact velocity of the erosive particle. If the speed is very low then stresses at impact are insufficient for plastic deformation to occur and much slower wear proceeds by surface fatigue. When the speed is elevated to a certain threshold value, of about 20 m/s [19], it is possible for the eroded material to deform plastically on particle impact. In

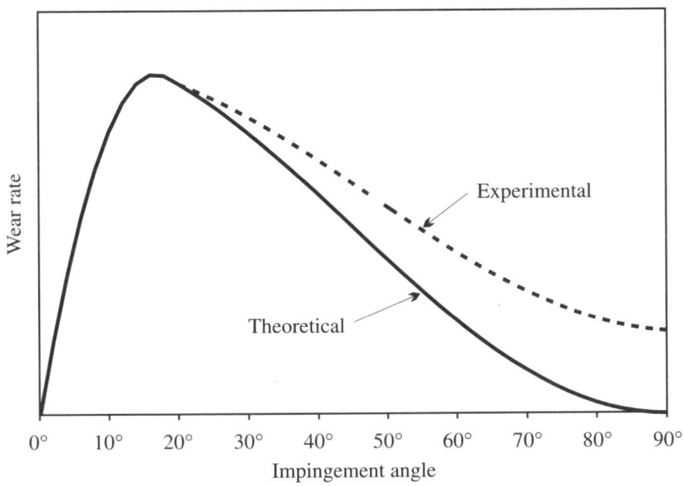

Figure 7.6 Variation of erosive wear rate with impingement angle [41].

this regime wear may occur by repetitive plastic deformation. For medium to high speeds the relationship between the erosive wear rate R_m, expressed as the volume removed by a given mass of abrasive, and impact velocity v can be described by a power law

$$R_m \approx kv^n \tag{7.5}$$

where k is the empirical constant and n the velocity exponent.

The early formulation has predicted the material removal rate proportional to the square of the impact velocity [44]. The experimental data, however, yields velocity exponents ranging between 2 and 3 [19,20,41].

In many diamond tool applications the use of water as a cooling liquid is widespread. Therefore, the matrix erosion occurs by the flowing action of a liquid stream, which carries solid workpiece particles. The characteristics of the liquid medium may have a strong effect on the final wear rate. It has been documented that small additions of lubricants to erosive slurries can markedly reduce wear [19]. Controlling factors relate to the bulk properties of the medium, i.e. viscosity, density and turbulence as well as to its microscopic properties such as corrosivity and lubrication capacity. The ability of the liquid to provide cooling during particle impingement is also important.

The drag forces imposed by a viscous slurry on erosive particles can affect wear by altering the impingement angle. The effect of the medium is assessed in terms of so-called 'collision efficiency', which is defined as the ratio of particles that actually strike the wearing surface to the total number of particles contained in the slurry jet that is directed at the target [19]. It has been found that the collision efficiency declines from a limiting value of 1.0 for large particles of 750 μm size to less than 0.1 for small particles of 75–90 μm size at medium dynamic viscosities [19]. The impact velocity is affected by the particle size in a similar manner and therefore larger particles strike the target at greater velocities than small particles do [45]. The reductions in collision efficiency and impact velocity are caused by the viscous medium which sweeps smaller particles past the wearing surface. The erosive wear rate is found to closely follow the same trend as collision efficiency and impact velocity, which indicates that the primary effect of a liquid medium is to draw the particles along the liquid streamlines and to divert them from the wearing surface. It is noteworthy that turbulence of the medium accelerates erosive wear since particle impingement is more likely to occur in the turbulent flow than in the laminar one.

In literature, the slurry erosion has received recognition as a mechanism that is likely to make contribution to the overall wear of diamond-impregnated tools [11,14–16,40,46]. Experimental evidence indicates, however, that the erosive wear mechanism has rather marginal importance and locates this form of wear, secondary to abrasion [18].

7.3 The Role of Diamond

During sawing and grinding the load is primarily borne by the cutting diamonds with only a small system dependent proportion carried by the swarf sandwiched between the workpiece and the matrix.

Diamond crystals protruding above the matrix aid its overall wear resistance by:

- preventing the matrix from direct rubbing against the processed material;
- direct blockage of abrasion grooves.

The former point extends to the complex issues of developing space (clearance) between the matrix and the surface of the processed material that would be adequate for the application needs.

The latter point is associated with a phenomenon termed as the 'stand-out effect' [19]. As illustrated in Fig. 7.7, inhibition of wear by the abrasive swarf leads to the development of a matrix tail and chip removal grooves behind each individual diamond particle, unless the working diamonds change orientation in relation to the direction of movement, as in the frame sawing of stone.

When pullout occurs the matrix tail is no longer shielded by the diamond crystal. Immediately after the particle loss it is likely that the tail is exposed to short but

Figure 7.7 Working face of circular (top) and frame (bottom) sawblade segments.

extremely severe friction directly against the workpiece. Hard asperities pass over its surface like a cutting tool, wearing the matrix many times faster than unfixed abrasive particles. Figure 7.8 shows a typical pullout site worn flat along the direction of the saw blade rotation.

The wear pattern on the elevated, former tail site displays a random topography suggesting gradual removal of surface layers by the successful contact of workpiece debris. The chip removal groove looks completely different. Due to wider clearance it has been polished smooth, by erosion, and shows no signs of abrasion.

Given that the crystal size is kept constant, increased diamond concentration in conjunction with improved retention capacity of the matrix yields more diamond particles on the wear surface. Therefore, due to the stand-out effect, the tool life should increase monotonically in proportion to the number of cutting points which, in the tool-making routine, is controlled by altering diamond size and concentration. The application practice indicates, however, that the sawblade life increases only until a certain optimum concentration is attained beyond which point the tool life stabilises [4] or even decreases [47].

That optimum diamond concentration is a complex function of its type and size, tool operating parameters, workpiece properties, machine condition, etc. Items requiring special attention are uniform distribution of diamond in the matrix and machine power capabilities.

With increasing the concentration a certain proportion of diamonds is shielded by other stones in front of them and is thereby rendered ineffective. It is therefore important that diamond crystals are uniformly distributed throughout the matrix.

The concentration at which maximum tool life is achieved is also dependent on the machine power ratings. Insufficient machine power do not guarantee adequate

Figure 7.8 Wear pattern around pullout site.

penetration of cutting points into the workpiece and thus may hinder self-sharpening of diamond particles by controlled micro-fracture [13]. In the extreme, extended wear flatting may result in no cutting action whatsoever.

At high rubbing speeds, which are customary in the use of diamond tools, and at very light loads a temperature of up to 500°C may be generated on the diamond faces which renders them blunt through oxidation or graphitisation/oxidation mechanism [48]. As it is virtually unrealistic that the cooling system fully compensates for the rise in temperature, the generated heat is partly transferred to the matrix thus making it softer. It has been documented theoretically [18] and experimentally [6] that normal forces acting on a sawblade grow steeply with progressive dulling of cutting points. Their detrimental effect on the tool life is twofold. First, the increasing normal force per working diamond promotes plastic deformation of the thermally softened matrix and may lead to excessive debonding and pullout of diamond particles; and second, unexpectedly high forces are hardly ever compensated for by the rigidity of the tool support. The tool thus becomes dynamically unbalanced which results in harmful vibrations generated in the system.

Due to great complexity and inherent instability of the actual wear conditions the current understanding of tool wear mechanisms remains highly qualitative. There are, however, clear experimental indications [18] that:

- On difficult to cut (high power-consuming) workpiece the diamond breakdown and/or pullout appear to be the pre-dominant forms of tool wear.
- On easy to cut (low power-consuming) workpiece the matrix is capable of retaining the diamond crystals longer and allow them to project much better beyond the working face of the tool. Hence, this is the matrix resistance to abrasive wear, which takes priority over capacity for diamond retention.

Assuming proper tool specification, to optimise its life it is necessary to avoid changes in the workpiece type and its machining conditions thus permitting development of a stable balance between the diamond particle protrusion and diamond penetration depth. Any alterations to the established cutting regime lead to accelerated tool wear by bringing the system closer to a new equilibrium.

References

1. Chalkley, J.R., Thomas, D.M., The tribological aspects of metal-bonded diamond grinding wheels. Powder Metallurgy, 1969, 12(24), 582–597.
2. Bailey, M.W., Collin, W.D., Investigations into diamond sawing using titanized grits. Stone Industries, 1977, 18–21.
3. Burgess, R.R., Circular sawing granite with diamond sawblades. Paper presented at Industrial Diamond Association of Japan 30th Anniversary Meeting and Seminar, Tokyo, Japan, May 16, 1978.
4. Burgess, R.R., Man-made diamond for stone processing. Paper presented at 1st Technical Symposium, Bucharest, Romania, October 5–6, 1978.
5. Bailey, M.W., Wright, D.N., SDA85 and SDA100 in circular saws – the effect of cutting parameters. In Proceedings of De Beers Düsseldorf Conference'79, Düsseldorf, Germany, May 22–23, 1979, pp. 2.15.1–2.15.12.

6. Mamalis, A.G., Schulze, R., Tönshoff, A.K., The slotting of blocks of hard rock with a diamond segmented circular sawblade. Industrial Diamond Review, 1979, 39(5), 356–365.
7. Ertingshausen, W., Wear processes in sawing hard stone. Industrial Diamond Review, 1985, 45(5), 254–258.
8. Wright, D.N., Wapler, H., Investigations and prediction of diamond wear when sawing. Annals of the CIRP, 1986, 35(1), 239–244.
9. Wright, D.N., The prediction of diamond wear in the sawing of stone. In Advances in Ultrahard Materials Application Technology, edited by Barrett, C., Vol. 4, De Beers Industrial Diamond Division, Ascot, 1988, pp. 47–60.
10. McGowan, J., Brauninger, G., The application of superabrasives in granite slabbing. Paper presented at Superabrasives'91 Conference, Chicago, USA, June 11–13, 1991.
11. Liao, Y.S., Luo, S.Y., Wear characteristics of sintered diamond composite during circular sawing. Wear, 1992, 157, 325–337.
12. Liao, Y.S., Luo, S.Y., Effects of matrix characteristics on diamond composites. Journal of Materials Science, 1993, 28, 1245–1251.
13. Davis, P.R., Fish, M.L., Peacock, S., Wright, D.N., An indicator system for saw grit. Industrial Diamond Review, 1996, 56(3), 78–87.
14. Tönshoff, A.K., Asche, J., Wear of metal-bond diamond tools in the machining of stone. Industrial Diamond Review, 1997, 57(1), 7–13.
15. Konstanty, J., Developing a better understanding of the bonding and wear mechanisms involved in using diamond impregnated tools. In Proceedings of International Workshop on Diamond Tool Production, Turin, Italy, November 8–10, 1999, pp. 97–106.
16. Konstanty, J., Diamond bonding and matrix wear mechanisms involved in circular sawing of stone. Industrial Diamond Review, 2000, 60(1), 55–65.
17. Romański, A., Doctoral Thesis, AGH, Kraków, 2000.
18. Konstanty, J., Cobalt as a Matrix in Diamond Impregnated Tools for Stone Sawing Applications, 2nd Edition. AGH Uczelniane Wydawnictwa Naukowo-Dydaktyczne, Krakow, 2003.
19. Stachowiak, G.W., Batchelor, A.W., Engineering Tribology. Tribology series 24, Elsevier Amsterdam–London–New York–Tokyo, 1993.
20. Sundararajan, G., The solid particle erosion of metallic materials: the rationalization of the influence of material variables. Wear, 1995, 186–187, 129–144.
21. Khruschov, M.M., Principles of abrasive wear. Wear, 1974, 28, 69–88.
22. Rabinowicz, E., Friction and Wear of Materials. Wiley, New York–London–Sydney, 1965.
23. Rabinowicz, E., Dunn, L.A., Russell, P.G., A study of abrasive wear under three-body conditions. Wear, 1961, 4, 345–355.
24. Rabinowicz, E., Mutis, A., Effect of abrasive particle size on wear. Wear, 1965, 8, 381–390.
25. Misra, A., Finnie, I., Correlations between two-body and three-body abrasion and erosion of metals. Wear, 1981, 68, 33–39.
26. Hurricks, P.L., Some metallurgical factors controlling the adhesive and abrasive wear resistance of steels. A review. Wear, 1973, 26, 285–304.
27. Uetz, U., Föhl, J., Wear as an energy transformation process. Wear, 1978, 49, 253–264.
28. Fang, L., Kong, X.L., Su, J.Y., Zhou Q.D., Movement patterns of abrasive particles in three-body abrasion. Wear, 1993, 162–164, 782–789.
29. Borik, F., Testing for abrasive wear. In Selection and Use of Wear Tests for Materials, edited by Bayer, R.G., ASTM STP 615, Philadelphia, USA, 1976, pp. 30–44.
30. Tucker, R.C., Miller, A.E., Low stress abrasive and adhesive wear testing. In Selection and Use of Wear Tests for Materials, edited by Bayer, R.G., ASTM STP 615, Philadelphia, USA, 1976, pp. 68–89.
31. Misra, A., Finnie, I., A classification of three-body abrasive wear and design of a new tester. Wear, 1980, 60, 111–121.
32. Ulvensøen, J.H., Müller, K., A new and flexible test-method for abrasive wear. In Proceedings of 12th International Plansee Seminar '89, Wear Resistant Materials, Reutte, Austria, May 8–12, 1989, Vol. 2, pp. 463–480.

33. Muller, K., Fundal, E., The Struers micro wear model and test. In Proceedings of 12th International Plansee Seminar '89, Wear Resistant Materials, Reutte, Austria, May 8–12, 1989, Vol. 2, pp. 481–495.
34. Fundal, E., The Struers micro wear test. Structure, 1989, 3, 3–4.
35. Micro Wear Test. Instruction Manual. Struers Tech, Rødovre/Copenhagen, Denmark, 1989.
36. Labormat – Basic concepts and machine design. Fundal Consulting, Borup, Denmark.
37. Application of the Labormat lapping and micro wear testing. Fundal Consulting, Borup, Denmark.
38. Standard test method for measuring abrasion using the dry sand/rubber wheel apparatus, ASTM Designation: G 65-94.
39. Hawk, J.A., Wilson, R.D., Tylczak, J.H., Doğan, Ö.N., Laboratory abrasive wear tests: investigation of test methods and alloy correlation. Wear, 1999, 225–229, 1031–1042.
40. Cai, O., Diamond segment production technology. Part five. Diamante Applicazioni & Tecnologia, 2003, 8(32), 51–59.
41. Finnie, I., Some reflections on the past and future of erosion. Wear, 1995, 186–187, 1–10.
42. Hutchings, I.M., Strain rate effects in microparticle impact. Journal of Physics D: Applied Physics, 1977, 10(14), L179–L184.
43. Talia, M., Lankarani, H., Talia, J.E., New experimental technique for the study and analysis of solid particle erosion mechanisms. Wear, 1999, 225–229, 1070–1077.
44. Finnie, I., Erosion of surfaces by solid particles. Wear, 1960, 3, 87–103.
45. Clark, H. McI., The influence of the flow field in slurry erosion. Wear, 1992, 152, 223–240.
46. Guangzhi, L. (ed.) Diamond Drilling Handbook. Geological Publishing House, Beijing, China, 1992, p. 129.
47. Reed, J.R., Applying metal bond diamond blades – art or science. Industrial Diamond Review, 1973, 33(4), 291–296.
48. Jones, W.D., Fundamental Principles of Powder Metallurgy. Edward Arnold Publishers Ltd., London, 1960, pp. 806–819.

CHAPTER 8

Main Application Areas and Operating Guidelines for Diamond-Impregnated Tools

The methods of making diamond-impregnated cutting tools have undergone a fantastic development since the 'invention' of synthetic diamond in the 1950s. Over the last five decades different types of metal-bonded tools have been increasingly used for the most demanding sawing, drilling, grinding and polishing operations to bring major productivity benefits in the processing of natural stone and construction materials, production of glass and ceramics, exploration of petroleum and gas, etc.

8.1 Sawing

Diamond-impregnated tools are now accepted as the most effective means of sawing natural stone, concrete, reinforced concrete, asphalt, brickwork, glass and other ceramic materials. The main classification of the operations carried out on these materials comprises circular- , frame- and wire sawing.

8.1.1 Circular Sawing

A diamond circular sawblade is an efficient and versatile tool, which may be used on both portable and stationary machines. Continuous rim blades and segmental laser welded blades are primarily designed for dry cutting operations carried out by means of a hand-held equipment although the latter are also recommended for wet sawing of asphalt and concrete by a floor saw machinery that may not ensure adequate water supply (Figs. 8.1 and 8.2).

Figure 8.1 Different rim configurations of dry cutting blades.

Figure 8.2 Floor saw fitted with 300 mm laser welded dry cutting blade.

Brazed segmental blades are commonly used for sawing walls (Fig. 8.3) and on stationary machines (Fig. 8.4), which secure sufficient amounts of coolant to prevent the brazed joint from overheating. These tools can be subjected to a number of re-tipping operations, which is economically justified in the case of bigger sawblades (see Fig. 8.4b). The contribution of the steel centres to the overall cost of the tool increases dramatically with the blade diameter and therefore it is reused until its fatigue life is approached.

Figure 8.3 Remotely controlled wall saw cutting a doorway in a concrete structure. (Courtesy of Hilti Corporation)

(a) (b)

Figure 8.4 Sawing stone on (a) a bridge saw and (b) 2.7 m diameter block saw.

Careful storage, handling, mounting and selection of application conditions [1] prolong the sawblade life and improve its performance. The very general guidelines for operating parameters used in the most typical circular sawblade applications are provided in Tables 8.1 and 8.2.

The easy to process, low power-consuming materials are generally sawn 'to full depth' with the downward rotation of the blade; whereas the difficult to cut, high power-consuming materials, such as granite or hard and dense ceramics, are always sawn in many passes with a reciprocating movement of the sawblade, which alternately operates in the up-cutting and down-cutting modes. In the latter case the

Table 8.1 General recommendations of peripheral speeds and cutting rates for use on single-blade machines.

Workpiece material	Blade peripheral speed (m/s)	Cutting rate (cm^2/min)	Comments
Dry cutting of various materials with hand-held equipment	80–100	Must be naturally 'accepted' by the sawblade	Tilting or twisting of the sawblade must be avoided
Quartziferous granites	25–30	100–200	Cutting rate grows with the machine power rating. Lower cutting rates are recommended whenever good surface finish is a priority
Low quartziferous granites	30–40	200–600	
Marbles	40–50	600–1200	
Travertines	45–60	800–1200	
Sandstones	40–65	300–1000	
Ceramics	20–50	—	
Concrete	35–50	—	
Reinforced concrete	30–40	—	
Asphalt	40–60	—	

Table 8.2 General recommendations of machine power ratings and coolant supply for sawing natural stone on stationary single-blade machines.

Sawblade diameter (mm)	Desirable power (kW)		Minimum water flow (l/min)
	Granites	Marbles, travertines, sandstones	
300–400	4	5	10
500	9	7	15
600	13	9	20
700	18	13	30
800–900	18	18	30
1000	22	22	40
1200	29	29	50
1400–1600	40	44	60
2000	44	48	70
2500	48	51	80
2700	55	59	80
3000	66	74	90

depth of cut ranges between 0.4 and 20 mm whereas the feed rate is chosen to comply with the recommended cutting rates (see Table 8.1).

High feed rates combined with shallow depths of cut will favour a free cutting action of the tool, which becomes the first priority in sawing granite on multi-blade machines for the production of modular tiles (Fig. 8.5). The latest generation of machines for this application is capable of using a spindle with up to 80 sawblades 1 m in diameter [2]. To ensure smooth running of the system, each of the vertical blades has to be matched within narrow specification in regard to ovality, balance, side run-out and tension, and fitted with sandwich-type segments containing high-grade diamond grits. Blocks are sawn at a feed rate of between 8 and 14 m/min and a depth of cut ranging between 0.4 and 1.5 mm [3].

The other important application conditions are given in Table 8.3.

Figure 8.5 Multi-blade saw cutting a granite block. (Courtesy of EHWA Diamond Ind. Co., Ltd.)

Table 8.3 General recommendations of machine power ratings and coolant supply for use on multi-blade machines [3].

Number of sawblades	Desirable power per blade (kW)	Sawblade diameter (mm)	Minimum water flow per blade (l/min)
5–9	55–66	1000	30
10–20	74–92	1100	35
21–32	110–132	1200	35
35–60	147–220	1600	40

8.1.2 Frame Sawing

Despite a comparatively high initial cost of machinery, frame saw systems have long been established as an efficient and low sawing cost means of mass production of stone slabs. They are generally used for cutting marble, travertine, limestone, sandstone and agglomerates, however, syenite and the easiest to process types of granite may also be sawn.

Frame sawblades are usually mounted horizontally on single or multi-blade (Fig. 8.6) machines, which are classified by the product of the length of stroke and the flywheel rotational speed (see Chapter 2). If the product is equal to or lower than 50000 mm/min the machines are regarded as 'slow frames', otherwise they are 'fast frames'.

The most typical sawing parameters used on slow and fast horizontal frames are listed in Tables 8.4 and 8.5.

It is noteworthy that the feed rates achievable on single-bladed frames may markedly exceed the figures given in Table 8.5 since, in general, the feed rate increases with the linear speed of the frame and decreases with the number of sawblades [6].

8.1.3 Wire Sawing

Over the past three decades, diamond wire sawing technology has made significant inroads into the stone industry, primarily for quarrying of stone blocks and their

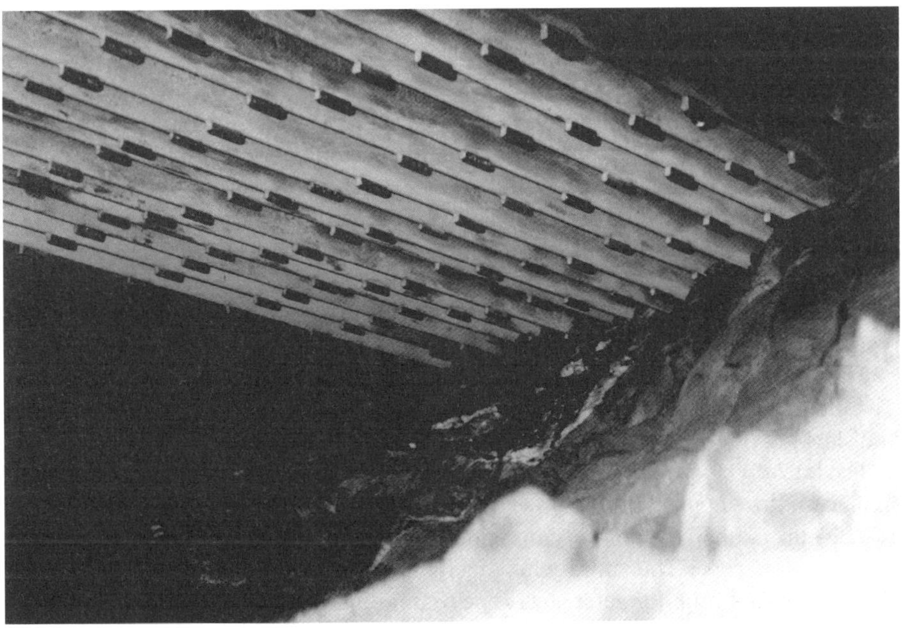

Figure 8.6 Frame saw, fitted with multiple diamond blades, in action.

Table 8.4 Recommended machine power ratings and amounts of coolant for sawing stone on slow and fast frame saws [4].

Number of segments per blade	Power per blade (kW)		Minimum water flow per blade (l/min)	
	Slow frame	Fast frame	Slow frame	Fast frame
20–30	0.8–1.1	1.1–1.6	7–8	9–10
31–40	1.1–1.4	1.5–1.9	7–8	9–10
>40	1.5–1.6	1.9–2.2	7–8	9–10

Table 8.5 Recommended cutting rates for sawing various types of stone on slow and fast frame saws [4,5].

Stone	Feed rate (cm/h)	
	Slow frame	Fast frame
Marble	10–18	20–35
Travertine	15–22	25–40
Limestone	12–25	20–30
Sandstone	20–30	30–40
Agglomerates	13–20	25–35
Easy to process types of granite/Syenite	8–18	—

further division, as well as into construction applications such as structure refurbishment and demolition operations.

There are several advantages to block extraction by the diamond-impregnated wire that have made this technique attractive to stone quarries (Fig. 8.7). It offers markedly higher production rates and is not so labour-intensive as the conventional methods. Additionally, it minimises waste in any case and dramatically increase output of blocks usable for monumental purposes in areas where flawed or fragile stone is quarried.

Typical wire saw characteristics and sawing conditions used in quarrying various types of natural stone are summarised in Table 8.6.

The diamond wire saw has also made rapid progress in the stoneyards, where single- and multi-wire stationary machines are increasingly used for block preparation and slabbing (Fig. 8.8), as well as for profiling of stone slabs.

The linear wire speeds and cutting rates achieved on the stationary machines are similar to those applied in the quarry. For most applications, the wire saw contains 10–11 mm diameter diamond-impregnated beads mounted on a 5 mm diameter stainless-steel rope. The multi-wire machines, however, utilise beads 7–8 mm in diameter on a 4 mm diameter steel rope to minimise kerf widths and thus to maximise the yield of slabs per block.

Figure 8.7 *Cutting a large block from the rock face. (Courtesy of EHWA Diamond Ind. Co., Ltd.)*

Table 8.6 Typical application parameters for quarrying stone by means of diamond-impregnated wire saws [1,7,8].

Type of stone	Number of beads per meter of wire length	Wire linear speed (m/s)	Cutting rate (m^2/h)
Quartziferous granites	37–40	19–23	1–3
Low quartziferous granites	30–39	23–28	3–5
Marble	27–30	30–40	6–8
Limestone	27–30	28–40	7–8
Travertine	27–30	40–45	8–9

Figure 8.8 Stationary diamond wire saw for squaring stone blocks. (Courtesy of MC DIAM Sp. z o.o.)

The versatility and economic advantages of the wire saw technology have also been recognised in the construction industry, where portable wire saw machines are used for various construction, renovation and controlled demolition purposes (Figs. 8.9 and 8.10). The ability of the diamond-impregnated wire to cut cleanly, quickly and accurately, with little noise and vibration, makes this tool an ideal alternative to blasting or jack hammering with flame cutting of the rebar, which were previously used in removal of thick sections of reinforced concrete or brickwork. The cutting rates achievable on various construction materials vary widely from 1–6 m^2/h on reinforced concrete, through 5–11 m^2/h on plain concrete, up to 10–18 m^2/h on masonry depending on the type of concrete aggregates, percentage of steel reinforcing and brick composition [9].

8.2 Drilling

Diamond-impregnated drills are efficient tools for cutting isolated holes in ornamental stone, concrete, asphalt, brickwork, glass and other non-metallic materials, stitch-drilling doorways and vents in heavily reinforced concrete structures, as well as for exploratory drilling in earth formations, which seems to belong to the oldest civil engineering applications of diamond. Depending on the application conditions

Figure 8.9 *Portable wire machine making a vertical cut in a concrete wall. (Courtesy of Hilti Corporation)*

Figure 8.10 *Diamond wire saw cutting a reinforced concrete pile into small sections to facilitate its removal. (Courtesy of EHWA Diamond Ind. Co., Ltd.)*

drills can be designed by varying the configuration (see Fig. 8.11) and composition of the working part, as well as by altering the manufacturing route.

Diamond-impregnated core drills used for drilling shallow holes in construction materials (Fig. 8.12) are made by the hot pressing route. In wet drilling, it is common to inject water through the centre of the drill to cool the working part of the tool and to flush the cuttings out. As a general rule, small diameter drills, up to around 32 mm, have a continuous drill crown design with waterways incorporated in it, whereas bigger drills have diamond-impregnated segments brazed to the end of a core barrel and suitably spaced to facilitate efficient evacuation of the slurry.

Drills used for dry coring by means of hand-held equipment are guided by a cemented carbide-tipped drill mounted axially in a short barrel encompassing lateral air-vents. The central guiding drill projects out over the segments, which, for safety reasons, should be laser welded to the barrel.

Typical core drilling parameters are listed in Tables 8.7 and 8.8.

Drill bits used in geological exploration and mining of mineral deposits (Fig. 8.13) are made by the infiltration route. The tool range includes coring and non-coring surface set diamond bits, as well as diamond-impregnated core bits. The surface set bits offer higher penetration rates but they are prone to damage, while drilling through broken and fractured formations. More frequently used are impregnated core bits, which are more robust and capable of achieving longer bit life. As compared with surface set bits, they display better directional stability and drilling

Figure 8.11 Selection of core drills used for drilling in stone, concrete and ceramics.

Figure 8.12 Portable drill rig mounted for drilling through a concrete wall. (Courtesy of Hilti Corporation)

Table 8.7 General recommendations of peripheral speeds and feed rates for core drilling stone, concrete and glass.

Workpiece material	Drill peripheral speed (m/s)	Feed rate (cm/min)
Very hard and difficult to cut ceramics	~1	1–2
Granite, difficult to cut types of stone	1–2	3–4
Glass	1.5–2.5	3–4
Concrete, reinforced concrete	2–4	4–8
Marble, travertine, limestone	3–5	5–10
Sandstone, asphalt	6–8	10–20

Table 8.8 General recommendations of machine power ratings and water supply for core drilling.

Core drill diameter (mm)	Machine power (kW)	Water flow (l/min)
<50	0.7–1.1	~2
50–100	1.0–2.3	2–4
100–250	2.0–3.5	4–8
250–500	3.5–7.5	8–12

performance in the hardest rock formations, enabling extraction of clean core for laboratory examination of rock structures even from broken strata.

To optimise the economy of drilling, the rotational speed and feed rate of an impregnated bit should be balanced to suit the rock type being drilled. These two

Figure 8.13 Impregnated core bits used for geological exploration. (Courtesy of Asahi Diamond Industrial Australia Pty Ltd.)

parameters combine to yield a bit revolutions per centimetre of penetration (RPC) index that has been found as an extremely useful guide to successful drilling. Ideally, the load on the bit should be adjusted to aim at 80–100 RPC to keep the bit sharp [10,11]. If the load is insufficient, the diamonds will polish and the bit will 'close up'. If the load is too high, excessive and uneven wear can occur. As a general rule, the pressure on the bit should always be kept below 150 kG/cm^2 [10]. If problems appear in adjustment of the rotational speed and load on the bit to bring the RPC index into the optimum range alterations to the composition of the bit crown are usually needed.

Another important parameter that affects the operating efficiency of the bit is the coolant flow. Usually, coolant flow rates higher by 20–30% are required when drilling with diamond-impregnated bits as compared to drilling with their surface set counterparts [10]. In most situations water is used as the coolant but various cutting oils or lubricating fluids [12] may also be used, especially when drilling hard rock formations [11].

Table 8.9 gives broad guidance on the amount of coolant required for impregnated core bits of different dimensions.

Table 8.9 General recommendations of coolant flow rates for impregnated core bits [11].

Bit size designation	Nominal hole diameter (mm)	Coolant flow (l/min)
A	48	11–15
B	60	23–30
N	75.7	30–38
H	96	38–45
P	122.6	68–87

8.3 Grinding and Polishing

In addition to sawing and drilling, metal-bonded diamond-impregnated tools have also been used with great success in a variety of grinding and polishing operations, offering the best combination of long tool life, shape retention of the working face, fast stock removal and good surface finish. Due to the numberless combinations of application requirements, tool and machine designs, it is beyond the bounds of possibility to tabulate the application guidelines, as it has been done in the preceding sections, within the limited scope of this publication.

In stone working, the calibrating, grinding and rough polishing steps are effectively realised with metal-bonded diamond tools installed on either single-head machines or multi-head automatic surface polishing lines. The latter are universally used in the manufacture of accurately sized stone tiles for flooring and wall cladding in public and private buildings.

The recent technological innovation has led to an enormous increase in the production of natural stone components used for interior decoration, such as kitchen worktops, tabletops, side panels for built-in furniture, staircase treads, window sills, door frames and many other fittings, which are normally made of wood. With this type of production, many different profile and edge finishes are required. They can economically be accomplished on numerically controlled contouring machines, fitted with a wide range of diamond tooling to cover all these possibilities (Figs. 8.14 and 8.15). The recent generation of contouring CNC machines, equipped with large tool stores, inclinable spindles, laser tool travel controls, touch probes to measure the workpiece dimensions, as well as user-friendly programming facilities, enable the most complicated stone-shaping operations to be carried out unattended.

Over recent decades, diamond-grinding techniques have also branched into civil engineering projects. Planing concrete floors and walls (Fig. 8.16), slitting concrete and brickwork structures for wiring (Fig. 8.17), grooving the surface of roads, bridges and airport runways, to reduce the risk of aquaplaning in wet weather, are typical examples of jobs carried out with metal-bonded diamond-impregnated tools.

Another important application of diamond-impregnated tools is in grinding glass and ceramic materials. Various shapes of metal-bonded cup wheels, peripheral wheels and pencil edge wheels are used here, to complement electroplated and resin-bonded tools, for bevelling of plate glass, shaping, edging and surfacing of architectural, furniture, automotive and optical glass components, artistic decoration of

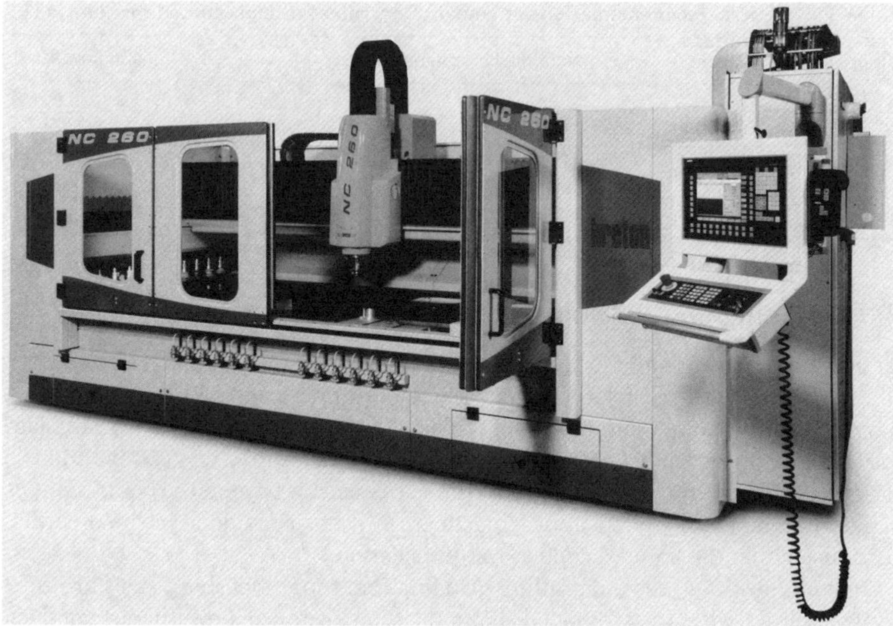

Figure 8.14 NC260 Contourbreton machine capable of handling very large stone slabs. (Courtesy of Breton S.p.A.)

Figure 8.15 Examples of metal-bonded diamond grinding tools for stone working.

(a)

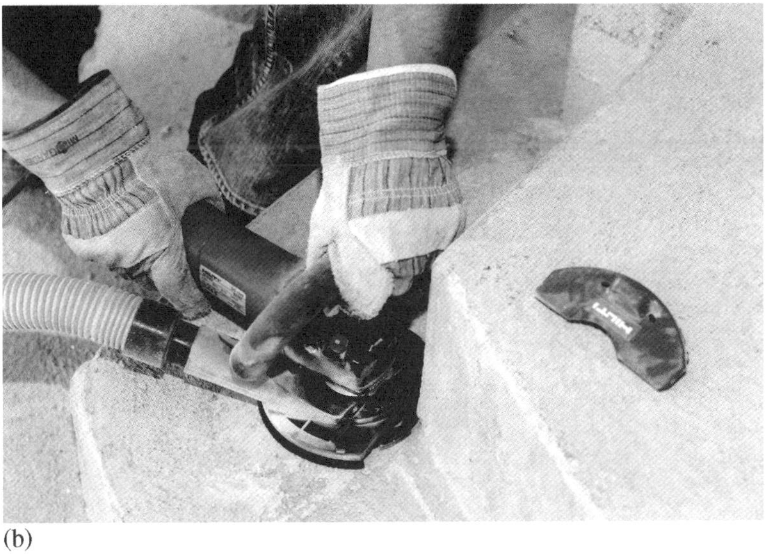

(b)

Figure 8.16 Clean grinding of (a) concrete face and (b) angle grinding of concrete steps. (Courtesy of Hilti Corporation)

glassware, etc. The grinding operations are carried out in several stages, mostly with metal-bonded wheels containing progressively finer grits, and require an adequate amount of coolant directed into the contact zone.

Modern manufacturing technologies are increasingly applying ceramics in the quest for materials which can withstand new operational extremes. As a result,

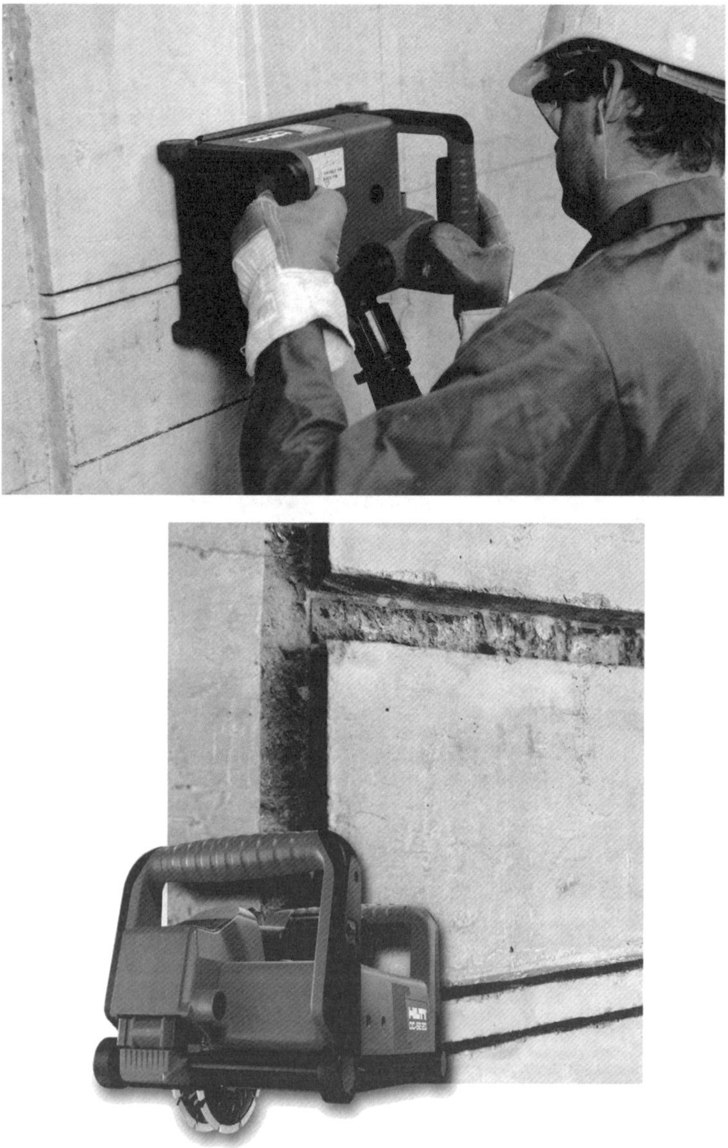

Figure 8.17 Two-bladed diamond slitter used to produce wider channels in concrete. (Courtesy of Hilti Corporation)

various types of ceramics have found broad application in electrical engineering (insulators, spark plug bodies, resistor supports, capacitors), construction of industrial furnaces (refractory bricks), production of accessories for laboratory use (tubes, crucibles, mortars, combustion boats), building and civil engineering projects (ceramic brickwork, roof tiles, piping), engineering of vehicles (motor parts), and as a host of other essential items of commercial and industrial equipment.

The fired ceramic components hardly ever have the required dimensions and, therefore, they have to be produced oversize for the exact dimensions and surface finish to be achieved by mechanical machining.

Without exception, ceramics and other non-metallic materials are hard and brittle and only diamond tools can grind, and polish, them easily and economically to the highest degree of precision and surface quality. As a general rule, tough, dense and fine-grained materials are more difficult to grind than coarse-grained or porous ceramics. As with other materials, it is impractical to provide general recommendations as to the optimum selection of grinding conditions, since the workpiece machinability, its shape, surface and dimensional specifications will depend on the characteristics of the material but also on the design and rigidity of the machine. Additionally, it has been found that the optimum machining parameters are strongly affected by factors such as firing temperature, crystallite size, density, type and distribution of additives and residual porosity. Consequently, the optimum combination of the diamond wheel composition and machining conditions must often be established empirically.

References

1. Parys, J.-M., Diamond tools for the stone industry. Diamand Boart S.A., 1995.
2. Davis, P.R., The future of diamond abrasives in stone processing. Industrial Diamond Review, 2001, 61(3), 159–167.
3. Anon., All what you absolutely need to know about multidisc machines. Diamand Boart S.A., 1995.
4. Anon., Specifications for the use of diamond frame saw blades. Diamant Boart.
5. Anon., Frame saw blades with diamond impregnated segments. Diamant Boart, 1L/1.77/2A.
6. Anon., Diamond tool types and uses. Directory '96 MmCLUB, pp. 84–108.
7. Hawkins, A.C., Antenen, A.P., Johnson, G., The diamond wire saw in quarrying granite and marble. Dimensional Stone, September 1990.
8. Diamond wire. MACO srl. Cantiano, Italy. Technical brochure.
9. Wright, D.N., Developments in diamond wire sawing of stone. Element Six internal paper.
10. Anon., Parameters for the use of diamond drill bits. Asahi Diamond Industrial Australia Pty, Ltd., A.C.N. 002 471 123.
11. Anon., Diamond Products Field Manual. Customer Guide to the Selection and Field Use of Diamond Coring Products. System Catalogue, 4th Edition, Boart Longyear Inc., 1996.
12. Songran, T., Fluids for diamond drilling. In Diamond Drilling Handbook, edited by Guangzhi, L., Geological Publishing House, Beijing, China, 1992, pp. 514–543.

Subject Index